As a retired bioscientist I am aware of the scientific evidence of global warming, its causes and effects. Dr. Robinson, however, has done more than just present the facts in an easy to understand and precise way. He has aroused my interest in the moral, psychological, even religious aspects of this critical issue. This book is a clear and concise explanation of the science of climate change and the factors relevant to finding a solution. It is a valuable resource for individuals and groups with inquiring minds wanting to explore and discuss the many aspects related to this topic. This book is a thoughtfully reasoned, scholarly discussion and a heartfelt, passionate plea to save our planet for future generations.

David W. Kramer, Ph.D.
Asst. Prof. Emeritus of Evolution, Ecology, and Organismal Biology
The Ohio State University

GLOBAL WARMING: CAN IT BE STOPPED?

The Science, Psychology, and Morality of Climate Change

PAUL E. ROBINSON, Ph.D.

Copyright © 2020 Paul E. Robinson, Ph.D..

All rights reserved. No part of this book may be used or reproduced by any means, graphic, electronic, or mechanical, including photocopying, recording, taping or by any information storage retrieval system without the written permission of the author except in the case of brief quotations embodied in critical articles and reviews.

This book is a work of non-fiction. Unless otherwise noted, the author and the publisher make no explicit guarantees as to the accuracy of the information contained in this book and in some cases, names of people and places have been altered to protect their privacy.

Archway Publishing books may be ordered through booksellers or by contacting:

Archway Publishing
1663 Liberty Drive
Bloomington, IN 47403
www.archwaypublishing.com
844-669-3957

Because of the dynamic nature of the Internet, any web addresses or links contained in this book may have changed since publication and may no longer be valid. The views expressed in this work are solely those of the author and do not necessarily reflect the views of the publisher, and the publisher hereby disclaims any responsibility for them.

Any people depicted in stock imagery provided by Getty Images are models, and such images are being used for illustrative purposes only. Certain stock imagery © Getty Images.

Scripture taken from the King James Version of the Bible.

ISBN: 978-1-4808-9549-2 (sc)
ISBN: 978-1-4808-9550-8 (e)

Library of Congress Control Number: 2020917041

Print information available on the last page.

Archway Publishing rev. date: 10/8/2020

This book is dedicated to the most important people in my life: my late mother (Mary) and father (Frank), my wife of 58 years, Lana Jeane (Bowman) Robinson and her parents, Joe and Elizabeth Bowman, all who love(d) me unconditionally; my sons, John and Nathan, and our daughters, Lisa and Tu, who have taught me so much about life and how to love fully; my grandchildren, Quinn, Caleb Oberrath, Aidan, and Cameron who enrich my life so much; and my brothers and sisters whose love and support sustained me over the years: Frank, Jerry, Patty Stanley, Roger, Dana McGraner, Dick, Dorothy Daugherty, Judy Jones, Norma Bails and Linda Jones.

Contents

Acknowledgements ix
Foreword xi
Preface xv
Introduction xxvii

Chapter 1   There Is A War Going On 1
Chapter 2   The Hard Facts 15
Chapter 3   Not All Beliefs Are Equal 35
Chapter 4   Why Facts Don't Matter 41
Chapter 5   Where Are Abraham's Children? 57
Chapter 6   There Is No Place Like Home 68
Chapter 7   For the Love of Our Mother Earth 78
Chapter 8   Can Global Warming Be Stopped? 84
Chapter 9   The Beginning or the End? 107
Chapter 10 Hopefully 116

Appendix A 121
Trusted Resources 123
Bibliography 125

Acknowledgements

My dear friend Jim Kulig believed in my ability to write this book and helped make me a better writer. His encouragement gave me the confidence to tackle this important and challenging task. I also appreciate the editorial assistance of Nelson Shogren. Dr. Paul Sukys deserves credit for his helpful critique. I also appreciate the support of Dr. David Kramer and the staff at Archway Pub. Co.

Foreword

"When written in Chinese, the word crisis is composed of two characters — one represents danger, and the other represents opportunity."

--- John F. Kennedy

One hundred years ago, Marshall Dawson wrote a controversial treatise titled, *Nineteenth Century Evolution and After*. In his Preface, Dawson makes the following observation, "At the present stage in human evolution, further progress depends upon what takes place in the mind, rather than upon changes in the thermometer and flour bin."[1] The same line might have been used by Paul Robinson in his book, *Global Warming: Can It be Stopped*, because that simple sentence expresses the entire premise upon which his book is written. Robinson believes, and he helps his readers believe, that the most daunting challenge facing us today is saving the environment and that the most demanding part of that crusade is changing people's minds about the nature of the challenge. More specifically, dealing with climate change requires that thousands, perhaps millions, of folks open their minds to the global catastrophe that awaits our children and grandchildren should we choose to do nothing.

Paul Robinson is neither a climate scientist nor a politician. He has no hidden agenda; no election to win; he has no contributors to woo to support a pet science project and no constituents to entice to support his

---

[1] Marshall Dawson, *Nineteenth Century Evolution and After: A Study of Personal Forces Affecting the Social Process in the Light of the Life-Sciences and Religion* (New York: The MacMillan Company, 1923), p. x.

re-election. Nor is he a zealot tilting at windmills and saving imaginary damsels in distress. He is not even a reporter chasing a Pulitzer Prize or a bureaucrat looking for a promotion. No, Paul Robinson is none of these things. He is, instead, simply an upright, honorable, well-informed man with a deeply held conviction that this world needs a voice, not the voice of panic, nor the voice of doom, nor the voice of anger. Instead what is needed is the voice of concern, the voice of hope, and the voice of reason.

In *Global Warming: Can It be Stopped* the reader will meet the calm, but firm voice of a man who, like most of us, is concerned about the future. Robinson has had a world of experience in human psychology. He has been a psychologist for over fifty years. During that time he has witnessed all levels of human suffering, cruelty, anger, hope, and longing. He has done his best to comfort and heal those in his charge. Now he has accepted a new client. No, that client is not the planet Earth. That would be too simple and too melodramatic. The Earth itself will heal and will survive. In fact, if the astrophysicists are correct, the Earth will live for another 5 billion years or so. Rather, the client is the human race, or more precisely, the human future. In *Global Warming: Can It be Stopped* Robinson explains how well-intentioned people can dodge, distort, and deny a reality that can no longer be ignored. That reality is not only that the Earth is warming at an alarming rate, but also that we have a responsibility to acknowledge that problem and to do something about it.

To do this Robinson takes the reader on a journey through the climate change crisis. He begins by exposing the reader to what he calls the "hard facts" of climate change. In a cleverly designed, approachable question-and-answer format, Robinson runs through virtually every question that a conscientious lay person might ask about climate change. Interestingly enough, Robinson's approach is not just aimed at true believers, that is, those convinced of climate change and committed to doing something about it. Nor does he simply target climate change cynics, those hardliners who deny the existence of climate change and who demonize those who disagree with them. Instead, he wisely focuses on those of us in the middle, those of us who are slowly awakening to the reality of climate change and are hungry for factual information that will satisfy our curiosity.

Robinson provides the reader with plenty of facts and he does so in a straightforward and understandable way. As readers stroll through Robinson's factual litany, they become aware that he is neither a technical wizard nor a bureaucratic record keeper but is, instead, a concerned, well-informed citizen with a desire for the reader to understand the facts as they exist now. Accordingly his language is as non-technical as possible, free of jargon, except when needed, and laid out in a conversational tone that communicates both the facts and Robinson's desire to ensure that the technical side of the problem is made clear to the novice.

From there he explores the nature of the scientific method and the difference between opinion and fact. Robinson is aware of the average person's skeptical attitude toward the dime-a-dozen talking heads who appear nightly on various news feeds and who spout facts and figures on both sides of an issue, leaving viewers bewildered and befuddled. Instead, Robinson provides a down-to-earth explanation of the difference between fact and belief and a series of uncomplicated and understandable examples that demonstrate, once and for all, that, in science there are right answers.

Robinson then moves on to what he does best: he explains the psychological reasons that people, sincere and authentic people, have difficulty accepting the reality of climate change. Robinson explains these psychological defense mechanisms (habits with fancy sounding names like, "cognitive dissonance," "confirmation bias," "emotional reasoning," and the "self-interest") in a way that projects both understanding and compassion, yet allows the reader to see the debilitating results of such behavior.

Had Robinson stopped here he would have produced a well-written booklet that covers both the regular fare of climate writers and his new psychological angle, and that might have been enough. However, Robinson does not stop here. Instead, he ventures into territory that is rarely covered by climate change writers, the spiritual and the moral side of the crisis. In a skillful way that both reports and challenges people of faith, Robinson explains how organized religion has joined the climate change crusade, while at the same time pointing out that they still have much more to do. In this way he obtains the support of those religious leaders and institutions involved in the fight and, at the same time, encourage others to follow suit. In a masterful stoke, he then adds a moral aspect to his polemic demonstrating that even the secular side of society has a responsibility to live up to their moral duty to protect the environment and to save the Earth for future generations. Robinson ends his treatise with a call to action and a series of steps that invite the reader to get off the fence, jump into the fray, and help slow or stop climate change.

To encourage dialogue among his readers, Robinson provides a series of discussion questions at the end of each chapter. His hope is that people will become more willing to discuss this critical topic with their family and friends. To help them do so he provides some guidelines to inform their discussions. He hopes this will prevent discussions from deteriorating into shouting matches or hurt feeling.

In sum, what we have here is a polemic that is not a polemic, a challenge that does not intimidate, and a call to action that is both plausible and doable to those of us who listen. If this is your first taste of Paul Robinson in action, then I envy you. You are in for a rollercoaster ride that will, in the end, convince you not only that climate change is real, but also that you will be morally and spiritually culpable for remaining on the fence.

Paul A. Sukys, J.D., Ph.D.
Professor Emeritus of Law, Literature, and Philosophy
North Central State College
Mansfield, Ohio
Author: *Lifting the Scientific Veil: Science Appreciation for the Nonscientist* (New York: Roman and Littlefield Publishers, Inc., 1999)

Preface

"You have stolen my dreams and my childhood with your empty words. And yet I am one of the lucky ones. People are suffering. People are dying. Entire ecosystems are collapsing."

"You are failing us. But the young people are starting to understand your betrayal. The eyes of future generations are upon you. And if you chose to fail us, I say: We will never forgive you."

-Greta Thunberg, Swedish activist

Before we focus on Greta's speech to the United Nations, allow me to admit that perhaps 2020 is not the best time to think about global warming/climate change. Americans, as well as the rest of the world, are preoccupied with the coronavirus or COVID- 19 pandemic. In addition, in the US we have a critical election coming this Fall when we decide to stick with the current occupant of the Whitehouse or elect a new person to guide our country through these difficult times. This election promises to be a highly contentious one. It may take months for the results to be finalized. Even then it is highly likely the country will be bitterly divided. Moreover, due to the coronavirus, many are deathly sick or dying while others are also unemployed or under-employed. Some have lost their homes or apartments while others are in danger of losing their homes or of not being able to pay their rent. Many struggle to provide for their basic needs such as health care, medicines, and food. To make matters even worse, we have to deny ourselves many of the pleasures of our former lives, including avoiding contact with loved ones and friends. In addition, racial and economic injustices continue to plague our nation as we have witnessed throughout this summer.

Meanwhile, people in other parts of the world are experiencing their own challenges and crises. Chronic hunger, starvation, droughts, floods, economic and political chaos, homelessness, wars and conflicts, displacement, death and destruction, lack of medical care, and so much more plague the lives of billions of people around the globe. In addition to all these challenges, they too are under threat of climate change.

The point of this entire litany of woes is that there is always a challenge or a crisis brewing somewhere in the world. We cannot wait, however, to address global warming/climate change until we have solved all the other problems facing humanity. There will never be an ideal, crisis-free, time to address the impending crisis of climate change. The world will never be free of problems and crises. Climate change must be addressed now. Just as the novel coronavirus required quick, decisive action, global warming demands the same response. If we delay our response to global warming (which by the way we are) the problem of climate change will only get worse as we discovered with our response to the coronavirus pandemic.

In short, these are difficult and, for many, perilous times. As the old saying goes, we have more problems than we can shake a stick at. Global warming is yet another challenge, one of the most important ones I would argue, that needs our immediate attention. As a psychologist, I am well aware that, as a group, our stress levels are quite high and that we have a limit to how much stress we can deal with at one time. The American Psychological Association has coined a new term, "ecoanxiety," to capture what the Association calls, "a chronic fear of environmental doom." We don't need to have a chronic fear of environmental doom, but we must be deeply concerned about the future health of our planet and, by extension, the health and wellbeing of every human being on this planet including, perhaps even more importantly, those yet unborn.

Meanwhile, as we fret about so many other things, the Earth continues to get hotter. Global warming is not going away even as we struggle with the many other challenges facing us and try to regain some semblance of a normal life. The coronavirus will probably be gone someday; not so global warming. Global warming is here to stay and will arguably be more enduring and damaging than the coronavirus in the long run. It will affect all humanity and its effects will be quite serious as well as life-threatening. Because global warming is seen more as a distance threat, it seems less urgent than our other stressors.

This book will make the case that it is not less urgent and that it is a problem we must face now. Decades ago scientists were warning about what was termed an "emerging virus" by a young virologist named Stephen Morse. Morse and other scientists were identifying climate change, massive urbanization, the proximity of farm and forest animals (virtual virus reservoirs) to humans, unrestricted international travel, and the movement of refugees due to famines and war as conditions that would make viral growth a certainty. If we had listened to these early warnings, perhaps we would not have been so affected by the COVID-19 pandemic. If we will only heed the warnings of climate scientists now, the effects of climate change can be mitigated.

If we listen to our climate scientists now who are warning about the dangers of global warming, we stand a good chance of arresting global warming before it creates more disruptions and dangers in our lives. Unlike SARS and MERS, virulent viral infections which were mostly confined to Asia and the Middle East, respectively, global warming is and will continue to affect every human being on this planet, much as COVID-19 is. There are at least three compelling reasons global warming is a problem that requires our immediate attention: national security, health, and economics.

In September 2009, the Central Intelligence Agency, CIA, launched a center called The Center on Climate Change and National Security. Its focus was to assess the national security impact of climate change phenomena such as desertification, rising sea levels, population shifts, and heightened competition for natural resources. The CIA recognized two realities: climate change is the number one threat to our national security and global warming is a problem we must deal with now. In 2019, the

Pentagon reaffirmed climate change as a national security threat a year after the President removed it from the administration's list of national threats (D'Angelo and Kaufman, *2019)*.

In addition, according to the National Oceanic and Atmospheric Administration, NOAA, climate change is a serious threat to our health. "Human health is vulnerable to climate change. The changing environment is expected to cause more heat stress, an increase in waterborne diseases, poor air quality, and diseases transmitted by insects and rodents. Extreme weather events can compound many of these health threats." Moreover, our food supply is threatened by increased temperatures, water stress, diseases, and weather extremes, all of which presents a major challenge to our farmers.

Climate change not only represents a danger to our national security and health and wellbeing, but it is also a drain on our national, state, and local budgets. For example, according to NOAA there were fourteen billion-dollar weather and climate events in 2018 and 2019. Losses were estimated to be $91 billion for 2018 and $45 billion for 2019. There has been nine consecutive years of eight or more billion-dollar disasters. According to NOAA, 2019 "was the fifth consecutive year in which 10 or more billion-dollar weather and climate events affected the United States." According to the government's own national climate assessment, continued warming "is expected to cause substantial net damage to the U.S. economy throughout this century, *especially in the absence of increased adaptation efforts (*italics mine)." It just makes good common sense as well as economic sense to address climate change as quickly as we can.

If we do not accept climate change as an urgent issue and address it effectively, we will have to pay the price in terms of increasing threats to our national security, health, and economy. We are already seeing the human toll of climate change as its life-changing effects are revealed through extreme weather events such as unprecedented heat waves, floods, droughts, and hurricane frequency and strength. Farmers struggle each year with planting and harvesting their crops due to unusual weather patterns. In addition, in terms of population shifts, Europe especially is experiencing the effects of mass migration, some of which is climate related. The same is happening in the U.S. as many migrate from Central America not only to escape rape, murder, and unemployment, but also to escape an increasingly hostile environment. Climate change as a critical issue can no longer be rationally denied or avoided.

We must ask if there will ever be a *good* time to address the emerging problem of climate change? And, if not now, when? Climate change is not going to go away just because we deny or ignore it. The health of our little planet, in fact, will only get worse as we delay taking effective action on it. We have seen this already in our delay in responding to the coronavirus. Our ignoring it only made matters worse. We must address climate change even as we struggle with other urgent matters too. We cannot wait until there is a good or better time. This is the exact time we should be addressing the problem of global warming.

We possess the intelligence and resources to deal with this urgent problem now. The real question is, do we have the will and wisdom to address this problem before it is "too late?" Scientists tell us we ignore the growing problem at our own peril. Our changing climate demands a robust and effective response on the part of all of us. After all, given that we all share the same little planet, we will have to work together to solve this mounting problem. This book is an attempt to inform people about the mounting problem of global warming and to motivate people to do something about it. It is not a book for academia, but for common, ordinary people who want to leave their children and grandchildren a healthy planet like the one we inherited from our ancestors. It's not too late for us to assure our loved ones have a planet that remains friendly to life. To paraphrase noted author and activist James Baldwin, not everything that is faced can be solved, but nothing can be solved until it is faced.

Therefore, let us start our examination of the critical issue of climate change/global warming by reflecting on the words of Greta. Greta has passion equaled only by her concerns for this planet and her generation. Sixteen year-old, Swedish environmental activist Greta Thunberg spoke the words above to the United Nations Action Summit in September 2019. Those who felt threatened or condemned by her words called her crazy and misguided. Others hailed her as an environmental hero, a champion for future generations as well as her own. Time magazine honored her by selecting her as 2019 Person of the Year. Her passion, commitment, and her willingness and ability to take a forceful stand inspire me to be a better steward and spokesperson for our planetary home. I appreciate Greta and believe she understands the seriousness of the impending climate crisis. Greta captures the true essence of the climate crisis by casting the debate as fundamentally a moral issue not a political or economic one.

It is courageous young people like Greta and a concern for future generations that prompted me to write about global warming. This book will make the case for human-caused or anthropogenic global warming. I am not a climatologist. I am, however, a student of science and one who has done his homework on this subject. I am a behavioral scientist which I believe uniquely qualifies me to write about the global warming crisis. So much of what drives the debate on global warming is psychology not physics or chemistry. It is beliefs, attitudes and even personality traits that make it a difficult subject for people. Much of the physics and chemistry of global warming are clear, some even indisputable. It is the psychological as well as religious factors that fuel the debate.

Unlike physical scientists, my laboratory is not in a confined space or the physical world, but the field of human behavior. I have always been curious about the how and why of things whether it be human behavior or the physical world. As a child I wanted to know about electricity, for example, so I tried an experiment that I definitely do *not* recommend anyone else try. I stuck a metal object into an electrical outlet! Oh, I found out about the nature and power of electricity really quick, and fortunately lived to tell about it.

I am drawn to all things science. I am curious. I want to know the how and why of things. The shape and color of leaves fascinates me. I have a passion for sea shells, rocks, and rock formations. Nothing is more pleasing to my eyes than the seemingly endless variety of shapes and colors of flowers. Is there a more awe-inspiring experience than to stand under the stars on a clear, dark night? Then there is our inner world- our fantastic and mysterious brain. I want to understand how that great inner space works, the workings of the human mind.

All this led me to major in teaching with a concentration in the sciences. As an undergraduate I studied many branches of science: astronomy, botany, chemistry, physics, geology, and zoology. Human behavior became my primary area of interest later which resulted in my becoming a psychologist. But over the years I continued to read about all things science.

Global warming and climate change captured my attention when it became clear that we were ruining our wonderful planet. I am writing because I actually do love our planet and all its inhabitants. I feel a moral responsibility as an inhabitant of this Earth and as a part of the human family to speak out about the way we are treating our celestial home and jeopardizing the future of our children and grandchildren- my children, and my grandchildren.

We are all part of the first generation of humans that have the power to alter Earth's environment forever. Our collective moral duty, therefore, is to save the Earth for future generations. When the industrial age started back in the late eighteenth century people did not know how the burning of fossil fuels would damage the Earth; but now we know and there is no excuse for not taking corrective action.

I take no comfort in the fact that at my age I am not likely to experience the worst of climate change. Nonetheless, my moral duty is to speak out and do all I can to save Earth from our collective neglect and abuse. I am convinced that it is morally wrong to continue to pollute our air, soil and water thus making generations to come pay the high price for our greed and neglect. We have no right to make them victims of our abuse of this planet anymore than a man has a right to abuse a woman making her the victim of his bad day or life or his drinking. Some day we will have to answer to our children. For me that day is today.

Because I care deeply about humanity, those born and not yet born, I devoted myself to learning more about climate change. I have researched the subject extensively. I am certainly not an expert in the field of climatology, but I have learned quite a bit through my study and have a working knowledge of the science behind it. I am confident that this book is an important addition to this issue of climate change as it not only addresses the science behind global warming, but also the relevance of psychology, religion, and spirituality to the issue.

This book is important and hopefully will be read not only by those already convinced of the anthropogenic nature of global warming but also by those who simply do not know what to believe and especially those who deny anthropogenic climate change. This book is important because critical decisions must be made regarding our response to global warming. These decisions need to be made now before the problem of global warming becomes worse. As with COVID 19, the sooner action is taken, the less harm and damage will be done to our Earth and, by extension, to humanity.

If we are to avoid the worst of climate change, it will depend on the actions of our legislators. They make the big decisions, governmental decisions, nation- and global-wide decisions, regarding global warming. They and they alone have the authority to enact legislation that can make a difference beyond what any one person or group of people can do. They can either pass laws that will help heal and prevent further damage to our environment or fail to do so thereby putting all of us at further risk of climate change. A public informed about the issues of climate change can bring pressure to bear on our legislators to do the right thing for our Earth and generations to come.

Not much will get done, as much is not currently getting done, to preserve our environment if the public sits on the fence uninformed and uninvolved. A well-informed public is our best offense and defense against a disinterested or even hostile legislative body as some legislators, doing the bidding of the fossil fuel industry, are. Scientists alone cannot convince legislators to make the right decisions regarding our environment. Change will come only when citizens join the efforts of scientists to press for legislation that will protect our planet from further harm. Information based on scientific facts, hence this book, is critical to informing the public and empowering the public to convince legislators to enact legislation to protect our planet.

Just how well informed is the American public about climate change? The Yale Program On Climate Change Communication conducted a scientific survey (2019) of a representative sample of the American people. Among its most relevant findings were these:

- 69% think global warming is happening while 16 % think it is not.
- 55% understand that global warming is mostly human caused, but 32% think it is due to natural changes in the earth's cycles.
- 62% say they are at least somewhat worried while 23% are very worried.
- 38% say they have personally experienced the effects of global warming.
- 36% say it is not personally important to them!
- 63% report that they rarely or never discuss global warming with family and friends.
- 9% consider it a religious issue while 38% consider it a moral issue.

Let's examine a few of these findings further. Still in this day and age one in approximately six Americans does not believe global warming is happening. That is disturbing. Equally disturbing is that about one in three do not consider it important to them personally. What, are they not thinking? Further, about one in three deny it is due to human activity. Fortunately, but also sadly, the vast majority are worried about it. As the data shows, however, most of us do not talk about it with those close to us. Like sex once was (believe it or not), it is a taboo subject, like politics and religion.

Yale and George Mason Universities have identified six Americas based on the level of concern for global warming. Those six Americas are: Dismissive (10%), Doubtful (10%), Disengaged (7%), Cautious (16%), Concerned (26%) and Alarmed (31%). These numbers show an increase in the number of Alarmed from 11% in 2014. Overall the level of Alarmed and Concerned rose from 45% to 57% in that same time period. Whereas, more and more Americans are expressing concern, still a large number are not showing realistic concern. Though the attitudes of Americans are changing, there is still much work to be done by all of us concerned about the growing threat of global warming.

Bishop John Shelby Spong (2000) writes about his background in the Southern culture of the 1940s and 1950s. He speaks of a time when the differences between races were obvious in so many ways--use of drinking fountains, toilets, public hotels, department stores, and in so many obvious and subtle ways. He writes: "It was a cruel system, and I was one of the unknown beneficiaries of it. Once again, I profited from this evil, *though at that point in my life I remained blissfully ignorant of it*" (italics mine).

Many of us are like Bishop Spong in that we, too, have profited from a system where fossil fuel use was just taken for granted. We did not know the damage we were inflicting on the environment. But now we know. Science has stripped us of our ignorance. If you read on, you will become convinced that we cannot go on as if burning fossil fuels does not matter anymore. We know better, and because we know better, we must do better. I hope this book will help us accomplish that.

I appeal to you the reader, for the sake of our children, grandchildren, and for the countless generations to come, to take this subject seriously. In global warming time, it is almost midnight. Time is of the essence. Scientists tell us we must act now before it is too late. There is no turning back the clock on global warming. Global warming is happening now and will only get worse the more we delay taking effective action. This book represents my attempt to speak for our children and grandchildren whose voice needs to be heard. I am also speaking for our planet that keeps sending us signals which we are largely ignoring.

If we all become informed and work together, we can preserve this wonderful planet for future generations. My firm belief is that we can turn the tide on global warming. Human beings have an amazing capacity for great achievements. We are the ones who have traveled distances our ancestors would never have dreamed possible. Whereas it would take months for our ancestors to travel from coast to coast, we can do it in a few hours. Very few of our ancestors ever envisioned us traveling to the moon and back. We are the people who have put the world's greatest libraries in the hands of anyone with a smart phone where data can be retrieved almost instantly. We have conquered diseases our ancestors never even understood. We have built tall buildings that put the Tower of Babel to shame.

As our mothers have said, "Where there's a will, there's a way." I believe that. I know we can accomplish almost anything when we really want to. My hope is that this book will inspire you to become a better steward of this wonderful, unique planet. My fondest desire is that all of humanity will have the will to correct the destructive course we are on and preserve the health of this planet for all grandchildren.

Allow me to close by quoting Anderson Cooper (2020), CNN Anchor, who spoke so eloquently about the recent birth of his son. He said: "I feel invested in the future in a way I hadn't really before. There's something about having a child that makes you feel connected to what is happening, and you want to make sure that the world this child is growing up in is a better one. You suddenly worry more about the future of all of us." It is my hope and prayer that every parent and grandparent, as well as every citizen of this Earth, shares the same concern for the future of our offspring and this amazing planet. Anderson said in a few words what I am trying to say about the reason I wrote this book.

Discussion Questions

1. What are your thoughts and feelings about Greta Thunberg's comments? Do her concern and opinions represent the youth you know? Your own? You can hear her speech on YouTube.

2. How concerned are you about global warming and climate change for yourself personally and for your children and their children?

3. Do you consider yourself sufficiently informed about the subjects of global warming and climate change? Do you know the difference between the two terms?

4. What is your reaction to the Yale survey?

5. Do you think we can and will prevent irreparable harm to our planet? How long will it take? What kind of Earth will our children inherit?

6. As you live your daily life do you think much about the affects of your activities on the planet and future generations?

7. What are your thoughts about Anderson Cooper's words regarding the future of his son?

Introduction

> Humpty Dumpty sat on a wall
> Humpty Dumpty had a great fall
> All the king's horses and all the king's men
> Couldn't put Humpty Dumpty together again.

The year was 1856. The occasion was the Eight Annual Meeting of the American Association for the Advancement of Science. The presenter was a woman who was not even a member of the Association, but she had a message which would reverberate down through the years. The paper, "Circumstances Affecting the Heat of the Sun's Rays," reporting on a simple experiment she had conducted, was only two pages long. Eunice Foote's conclusion was simple: "An atmosphere of that gas would give to our Earth a high temperature; and if as some propose, *at one period of its history the air had mixed with it a larger proportion than at present, an increase temperature from its own action*" (italics mine). The gas she was referring to was carbonic acid, H_2CO_3. She also found that the action of the sun is greater in moist air. Three years later, her results were repeated by John Tyndall.

What if her research had been taken more seriously and had spurred an investigation into the effects of CO_2 on Earth's temperature? The Industrial Revolution with its temperature increasing effect of CO_2 emissions was just a few decades old at that time. Actions taken then to reduce our carbon footprint could have saved us from the effects of climate change we are beginning to experience now. She was a woman ahead of her times, and, to her misfortune, ahead of the men of her time.

The above nursery rhyme serves as parable of our time. Scientists are warning us that there is a point of no return with regards to global warming. There is a tipping point- a point beyond which the concentration of carbon dioxide, CO_2, in the air is so great that the Earth becomes so hot and climate change is irreversible. The lesson of Humpty Dumpty is that some outcomes can be tragic and irreversible. Just as Humpty Dumpty could not be fixed, we are in danger of reaching a level of CO_2 in our atmosphere that will be impossible to recover from. Scientists warn that if we continue to release greenhouse gases, GHGs, into our atmosphere we will soon reach the point of no return, a point where it will take generations or centuries for them to dissipate or be absorbed.

As we debate the evidence of anthropogenic, human-caused, global warming and its probable effects on our weather and climate, we must face the reality that our Earth is fragile and, indeed, can be damaged beyond any quick fix. Scientists are unequivocal. There is overwhelming scientific agreement among climate scientists that our atmosphere is being dangerously altered by the emission of greenhouse gases. Scientists from all governmental organizations such as NASA (National Aeronautic and Space Administration) and NOAA (National Oceanic and Atmospheric Administration) as well as other public and private research institutes around the world all agree we are poisoning our air. The vast majority of scientists have concluded that if we do not stop releasing greenhouse gases into our atmosphere, we will not be able to keep our planet from warming beyond the critical point of 2.0 degrees Fahrenheit.

Global warming is described as an existential crisis precisely because it is a threat to our lives in so many ways. As Pogo famously said, "We have met the enemy and he is us." Global warming threatens where we live, how we heat and cool our homes, our reserves of fresh water, our agricultural production, our health, our economy and investments, and so many other aspects of our lives. Already many are directly experiencing the effects of global warming. The quality of life of generations to come is at stake. If global warming is not diminished, we will condemn future generations to struggle with an Earth that is no longer friendly to life. The effects of global warming are and will be more than an inconvenience.

Global warming is also a moral and spiritual crisis. It is a moral issue because our actions affect others. What one person does affects others. No one can hide from global warming any more than one can hide from COVID 19, the coronavirus. Moreover, global warming is a spiritual issue because global warming challenges our way of perceiving our world and being in relationship with others.

The intent of this book is to bring together the fields of climate science and psychology in the context of morality and spirituality. Global warming and our response to it are best understood in its larger context. Global warming is as much a psychological, moral and spiritual issue as it is a scientific one. To that end, this book has five desired outcomes.

The first is I hope the reader will come to a keener appreciation of the wonder, beauty and uniqueness of our little planet. Earth is truly a marvelous creation, and now we know fragile too. In fact, I hope you will come to love the Earth and its inhabitants as I do-- hopefully even more so. Greater love of this Earth and all its inhabitants is the only thing that will save this planet for future generations.

The second outcome I hope to achieve is to clear up some of the confusion about climate change and global warming. Make no mistake about it the science is in. The facts are clear; global warming is real and is caused by human behavior! Many, however, are confused about it-- and understandably so. They don't know who or what to believe. It is denied at the highest levels of government and the science can be confusing. This confusion, however, is dangerous because confusion results in inaction or poor decisions.

Climate change is not a matter of belief any more than the moon is a matter of belief. Like the moon, climate change is real. No one asks if you believe in the moon. No one should ask if you believe in global warming. Just like the moon can be seen, thereby proving its existence, so can human-caused global warming can be "seen" in the data scientists collect. The fact that the climate is changing and that it is due to human activity is no longer debatable. Some aspects of global warming are and they will be discussed later. One can have an opinion about global warming, as many do, but opinions are not facts. My hope is to clear up much of the confusion about what is known and not known about climate change. What to do about global warming can and should be debated, but it is a colossal and tragic waste of time to debate its principle cause.

To be clear, my task is not to convince anyone of anything about global warming but to get out of the way and let the evidence speak for itself. At times, I suspect my love and passion for this Earth and humanity may get in the way of the facts. For that I apologize. I simply want the reader to understand why virtually all climate scientists believe in anthropogenic, human caused, global warming. I am convinced that an objective consideration of the data will be enough to persuade the reader to the position taken by the vast majority of climate scientists.

This brings me to the third outcome I hope is achieved. My expectation is that people will be able to think more objectively, more clearly, about climate change. I will point out the opinions often substitute for facts in this debate. Still many people refuse to believe the facts about global warming and climate change. Some even seem to care less about the subject! This section will be devoted to understanding human behavior--my area of some expertise. Why and how people can deny facts or reality is a fascinating topic to me. I hope you will find it the same and that you will gain some insight into your own behavior as it relates to the denial and acceptance of facts. We will examine the psychological motivations and mechanisms that allow us to do so especially as they apply to global warming.

The fourth desired outcome is that the reader will think of global warming as a moral and spiritual issue. I will strongly assert that fundamentally climate change is not a political issue or a liberal vs. conservative issue. I will especially challenge people of faith to take a stand on this issue. Abraham's children, Christians, Jews and Muslims, as God-appointed stewards of His creation, will be encouraged to become the stewards their God intended them to be.

To believe it is a political issue and not a moral and spiritual issue is an intellectual and moral dodge. Making it a political issue confuses the public and allows those who deny climate change to avoid any moral and spiritual responsibility for climate change. Such belief also allows them to avoid doing anything about climate change and to go on with business as usual as Greta claims. We have long passed the point where we can afford to ignore or deny the facts. Way too much is at stake, if not for some of us because of our age, but certainly for generations to come.

The fifth outcome I hope to achieve is to motivate the reader to do something to help stem the tide of global warming. Climate change is affecting everyone or soon will be if it is not already; thus it is the responsibility of all of Earth's inhabitants to do something about it. Clearly some people bear more responsibility than others for its cause, but it is the moral duty of all of us to take climate change seriously and act accordingly, each in his or her own way. Just as with the 2019 novel coronavirus, it will take the collective action of people everywhere to stem the tide of climate change.

We will look at what is being done and what more can and needs to be done. Many are taking action. Individuals, groups, governments and corporations are acting to stem the tide of global warming even while the US government is attempting to negate many of the environmental rules and regulations designed to protect our environment and people. We must act collectively to stop the warming of our planet. We have no planet "B" to fall back on.

My hope is that the reader will become more informed, learn more about how they process information, be more understanding and tolerant of those who disagree with them, and become an activist for the climate and the health and wellbeing of all Earth's people, both those alive now and those not yet born. Climate change demands that we insist on the truth and that we relentlessly pursue it.

To promote discussion and dissemination of the facts of global warming, I have included a series of questions following each chapter for group discussion or discussion among family members and/or groups of friends. Many of the questions require openness and honesty. Some will find the questions more challenging than others. It is important that we strive to listen to the point of view of others, especially those with whom we disagree. It is imperative that these discussions not deteriorate into shouting matches where animosity is sown and the message gets lost. Please listen to your better nature when having discussions with others. The reader will find the suggestions for discussing the issue as detailed in Chapter Eight helpful.

Discussion Questions

1. Do you agree that the nursery rhyme Humpty Dumpty is a parable for our times in respect to global warming? Are there other rhymes or stories that serve as parables for our time, for example, the Three Little Pigs?

2. Do you agree or disagree that global warming is primarily caused by human activity? Or, as Pogo suggests, we are the enemy? What is the main evidence that human activity is the culprit?

3. How important are the author's five hopes for you personally?

4. The author states, "To believe it is a political issue and not a moral and spiritual issue is an intellectual and moral dodge." Do you agree or disagree?

1

There Is A War Going On

"Doubt is our product since it is the best means of competing with the 'body of facts' that exists in the minds of general public."

-unnamed tobacco executive (1969)

Dr. James Hansen (1988) was the first scientist to testify publicly in Congress about the dangers of human-caused global warming. He testified: "It is time to stop waffling… the greenhouse effect is here." From that momentous testimony, the battle was on. The Reagan administration had drafts of his congressional testimony repeatedly edited in order to weaken his controversial position, even controversial among scientists at that time. Hansen became the subject of so much criticism that a book was written about all the attacks on him. See Mark Bowen's book (2008) for more details about the assault on Dr. Hansen.

The opening quote from David Michaels' book (2008) captures the essence of the attack of the fossil fuel industry on the science of global warming. Naomi Oreskes and Erik Conway (2010) further detail the tactics of the industry to cast doubt on the science and the scientists. They charge the fossil fuel industry of creating a virtual Potemkin village of pseudoscience institutions including think tanks, journals, newsletters, and a core of supposed experts to attack the science and scientists. The latter actually includes a few scientists with impeccable credentials and outstanding reputations in their respective fields of science, that is, before they started receiving financial support from the fossil fuel industry.

The war is essentially a battle over the issue of our health on the one hand and the issue of the economy on the other or profit versus health. The fossil fuel industry recognizes that the public must continue to burn fossil fuels or they would go bankrupt. This same theme, profit versus health, is playing out now in respect to the coronavirus pandemic. There is much debate about when America should return to work. Our federal government has established guidelines, but some are ignoring those in their eagerness to get back to work and get our economy going again. Though most Americans support the science-based guidelines, many are choosing to ignore them. Business owners and corporations want America working and rightly so.

People want to work again. They need too. Many have no income at all right now. People have taken to the streets to demand their right to go back to work. Government checks are very slow in coming for many and some have not received any money at all. What monetary relief people have received is insufficient for most. Some are angry at their government and the scientists who are advising it. As fine and as respected a scientist as Dr. Fauci is, even he is under attack from fringe groups. They see him as an enemy of the people's rights to work and possess weapons! Many people also see climate scientists as "enemies of the people" too and are relentless in their attacks on them.

People need to get back to work to support their families and pay their bills while at the same time they need to protect their health- a most unfortunate dilemma. At this point in our country, people also need fossil fuels. There would be much suffering, not to mention hardship and perhaps economic ruin for those dependent on the fossil fuel industry, if suddenly we all stopped using fossil fuels. Though there is a slow transition from dependence on fossil fuels taking place, it is not fast enough to have much effect on our environment, on global warming. The balance between our health and the health of the planet on the one hand and our personal and national economy on the other is a delicate one to say the least. To prevent or reduce global warming, however, we must transition from fossil fuels as fast as we can.

Our concern for our health and the health of this planet and the need for economic security sets the stage for the battle the fossil fuel industry is waging against scientists who dare publish the fact that global warming is primarily caused by the use of fossil fuels. As Amy Westervelt made clear in a *Washington Post* article (2019), there is a "war of ideas" taking place in America and, we can say, around the world. This war pits the fossil fuel industry and its disciples against the best of our climate scientists. A war for the hearts and souls of all of humanity is ensuing. Truth and facts are at stake in this war as are profits and human welfare. More importantly, the fate of humanity is at stake!

No cost is being spared by the fossil fuel industry to maintain the status quo, i.e., to keep the oil wells pumping even as the Earth continues to overheat. Unfortunately, global warming has become a political and partisan issue. People have chosen sides not based on the evidence but based on discipleship to a political party.

Climate change has become an article of faith for many conservative Republicans and evangelicals. For example, global warming was called "a flat out lie" by an attendee at a Republican political forum in Indiana a few years ago. The man continued, "I read my Bible. He made this Earth for us to utilize." Another attendee at that meeting went on to say, "This so-called climate science is just ridiculous." The default position of those who deny the science of global warming is "the science is not settled." It is recognized that two people do not represent the entire view of the Republican Party; however, they do represent the view of many.

What should concern every American is the Trump administration's careless treatment of environmental issues in general and global warming in particular. Nadja Popovich and others (2019) detail how the current administration has tried to roll back nearly one hundred pieces of environmental legislation enacted by previous administrations. Fortunately, environmental advocate groups have had some successes in blocking these attempts in the courts.

Equally disturbing is a survey conducted by the Center for American

Progress Action Fund (1/28/19). The survey revealed that "there are 150 members of the 116th Congress--all Republicans--who do not believe in the scientific consensus that human activity is making the Earth's climate change." The good news is that is actually a slight reduction in the number of members who deny climate change.

Please understand. It is not my intent to single Republicans out as global warming deniers here though it may seem that way. Not all Republicans deny climate change any more than all Democrats believe in it. Besides, the word "Republican" means different things to different people. The labels "Republican" and "Democrat" are so imprecise that they have little meaning any more. It is impossible to group them into two distinct categories. No one person or group defines any party.

In addition, it was President Nixon who presented a 37-point message on the environment which called for the creation of the Environmental Protection Agency in 1970. The Clean Air Act was passed in 1970 to "foster the growth of a strong American economy while improving human health and environment. (EPA)" This Act enjoyed strong bipartisan support as did the revision in 1990 under Republican President, George H.W. Bush. Further, It was bipartisan support that lead to the passage of the Clean Water Act of 1972 which Nixon, interestingly enough, vetoed out of concern for the size of the Act's budget. Again, in both the House and the Senate, strong bipartisan support resulted in an override of Nixon's veto. Protection of the environment was once favored by both Republicans and Democrats.

Those who deny climate change, both Republican and Democrats, have been grouped into four categories by the Center for American Progress Action Fund. These are as follows and I quote:

1. Believing that climate change is not real or a hoax
2. Believing that the climate has always been changing and continues to do so, and saying that the Earth is just in a standard cycle of warming, despite evidence of faster change than ever before

3. Thinking that the science around climate change is not settled, or claiming that since they are not scientists themselves, they cannot know for certain
4. Believing that humans are contributing to a changing climate but are not the main contributors—again despite overwhelming evidence to the contrary

How can anyone continue to deny anthropogenic climate change? After all, it has to be obvious to everyone that we are polluting our atmosphere with greenhouse gases. The fact that we are has been known or at least speculated about for decades now. The only thing new is that our data is more accurate and comprehensive and our prediction models are more refined. As far back as 1965, President Lyndon Johnson was advised of the risks of continued carbon pollution. Being an oil man from Texas, he chose to ignore the warning. In 1969, President Nixon was also warned of the dangers of continuing to burn fossil fuels and thereby releasing more carbon dioxide into our atmosphere. Though Nixon showed more interest in environmental issues than any other previous president, he did little to address the burning of fossil fuels.

Even the fossil fuel industry knew of the dangers before the public began to become aware of the association between the burning of fossil fuels and the buildup of greenhouse gases, principally carbon dioxide, CO_2, in our atmosphere. An internal Exxon oil company memo (Union of Concerned Scientists, 2015) which came to light due to lawsuits against Exxon and other major oil companies, reveals that scientists working for them also warned of the dangers of burning fossil fuels as early as 1982. Memos warned of the threat of reducing our dependence on fossil fuels saying it would have severe impact on their global business and the economies of the world.

The most obvious conclusion from the three examples above is that, for a variety of reasons, mostly financial it seems, no one was willing to take the threats seriously enough to take appropriate action. Politicians didn't. Corporations didn't. The only group that was sufficiently concerned was the scientists who were looking at the growing and impending threat. Some of the public was aware but they were confused by the complexity of the data or by feuding scientists. And, of course, a misinformation campaign conducted by the fossil fuel industry, much like during the "tobacco wars" of the fifties, only confused the public more. This war if fueled by the fossil fuel industry whose motto seems to be, drill and spin (as in the facts).

John Broder (2010) wrote: "[T]he fossil fuel industry have for decades waged a concerted campaign to raise doubts about the science of global warming and to undermine policies to address it. They have created and lavishly financed institutes to produce anti-global-warming studies, paid for rallies and web sites to question the science, and generated scores of economic analyses that purport to show that policies to reduce emissions of climate-altering gases will have a devastating effect on jobs and the overall economy."

Amy Westervelt's essay (2019) is illustrative here. She identified several ways the industry has been successful in getting the public to believe that global warming is a scientific uncertainty. From influencing media outlets to report "uncertainties" to targeting conservatives with the idea that climate change is a liberal hoax, the fossil fuel industry has had success in confusing the public and lawmakers. In addition, it used its considerable influence and money to position contrarian scientists as experts as believable to the public as the more credible climate scientists.

Readers interested in learning more about the evidence for a well-organized, well-funded, and coordinated efforts to cast doubt on the science of global warming is referred to Michael Mann's book (2012) cites several examples of how the fossil fuel industry attacked climate science. Allow me to give you two examples from his book about how this war is being fought. This first example occurred in 1995 at a meeting of the Intergovernmental Agency on Climate Change (IPCC) meeting in Madrid, Spain. A fierce debate broke out between the scientists and delegates from the Mid-east oil countries. The battle was over a single word. The scientists wanted to report that the effect of human behavior on climate change was "appreciable." The Saudis, in particular, preferred a softer word, a word that was less accusatory. After two days of wrangling, they finally agreed on the word, "discernible."

Another example involved personal and professional attacks on a climate scientist, Ben Santer, a researcher with the Department of Energy's Livermore National Laboratory in California and the recipient of a prestigious McArthur "genius" award. Santer was the primary author on a series of papers establishing the role of human activity on climate change. Even though Hansen had first testified publicly about anthropogenic global warming, it was Santer's and his colleagues' work in the mid-1990s that provided the basic research that established the role of human activity in climate change.

Sander came under attack from a variety of sources. One was a scientist at the University of Virginia whose work was being subsidized by the fossil fuel industry. Another attack came from a group called the "Global Climate Coalition," a group also funded by the fossil fuel industry. This group accused Sander of abusing his power of peer review and of "political tampering" and "scientific cleansing."

Even the media joined in the attack. Dr. Frederick Seitz, founding chairman of another fossil fuel-industry-funded group, the George C. Marshall Institute, published an op-ed in the Wall Street Journal (1996) repeating the same charges of "political tampering" and "scientific cleansing." This latter charge was particularly malicious because Santer had lost relatives in Nazi Germany.

This misinformation campaign caused and is still causing confusion among the general public as well as our legislators. Such confusion is threatening not only the health of our planet but our health as well. Confusion delays or prevents action and works to the advantage of the fossil fuel industry. Since no one in a power position took the threat seriously enough or even denied it, nothing was done early on when much of the current threat could have been avoided. Even now global warming is not being taken seriously by those in a position to do the most about it. For example, President Trump tweeted: "The concept of global warming was created by and for the Chinese in order to make U.S. manufacturing noncompetitive." The denial of global warming is largely about money and the GNP, gross national product, for those who deny the scientific facts.

Paul Krugman (2020) writes that "zombie ideas" are ideas that have been proven wrong by overwhelming evidence and should be dead, but somehow keep "shambling" along, eating people's brains. He asserts that the most consequential "zombie" is climate denial and that it is kept alive by financial self interests. That certainly was and is today the motive of the fossil fuel industry. Upton Sinclair summarized the issue well when he wrote: "It is difficult to get a man to understand something when his salary depends on his not understanding it." For example, data from the Center for Responsive Politics (2019) revealed that the 150 Republican members of the 116th Congress accepted a total of $68 million in direct contributions from the fossil fuel industry. That is an average of $453,333 per member!

These kinds of "payoffs" make any possibility of joint congressional action on climate change extremely difficult. Why aren't more people worried and demanding action on the environment? Why don't more people see the importance of action now? Thomas Hegghammer's book (2017) about militant Islamist terrorism made this observation: "There is a difference between man-made and natural disasters. People are typically more afraid of man-made threats even if they are less damaging."

The number of deaths in the US due to the coronavirus pandemic exceeded the number of deaths due to the terrorist attack on 9/11. Clearly, we showed more fear of the coronavirus than we do of global warming even though global warming is predicted to cause unparalleled disruptions in our lives and also a high, if not higher, number of deaths. If we think the coronavirus virus was disruptive, wait until the worse for global warming hits us!

The reason for the greater fear of man-made threats has to do with the elements of predictability and control. Beyond our physical needs, such as air and water, for example, our psychological comfort requires that we be able to make a reasonable prediction about the future and believe we have some element of control over it. The unknown and unpredictable creates the greater amount of fear for us. We believe we have more control over natural threats than we do man-made ones; therefore, we fear natural events less. After all, natural threats, such as floods and tornadoes, are more or less predictable and often can be avoided, the element of control. On the other hand, a man-made threat, such as a terrorist attack, is not always predictable or controllable.

All these elements came into play with the coronavirus pandemic. We knew it was coming and had the opportunity to prepare for it. Even though we did not prepare as well as we should have, we, nonetheless, had the opportunity to mitigate its impact. If a terrorist attack, a man-made threat, had been predicted, the response of our government would have been swift and decisive to prevent it. Global warming is a natural threat and predictably but, unfortunately I must add, our government in Washington is treating it like it is not real, a hoax.

The war has been joined, unintentionally I must add, by many in the American public. For example, such disregard of a natural threat was evidenced over Spring break among adolescents when thousands of college students flooded the beaches in Florida. They not only were risking contracting the virus themselves, but also spreading it to other people. One reveler on a beach captured the attitude of many who deny global warming when she said during a live TV interview, "If I get corona, I get corona. At the end of the day, I'm not going to let it stop me from partying."

Far too many have adopted this same attitude toward the Earth. They will keep on partying regardless how hot the Earth is becoming. They will not let the facts of global warming convince them they have to make some adjustments in their lifestyles. Thinking we are invincible, that the damage to the Earth is minor or non-existent and that we are immune from natural disasters is foolish thinking. Denial of a problem does not make it go away. Denial delays any effective response to the problem. We must start taking the natural threat of global warming more seriously if we are to avoid a worst-case scenario.

Make no mistake about it. In spite of the fact that so many have ignored or minimized the threat of global warming, it is a serious problem that requires bold action. Such action is only beginning to be taken now some fifty plus years after the initial warnings. The Kyoto Protocol adopted in 1997 by many nations was a small step in the right direction. Unfortunately, one of the major players, the US, refused to adopt it. This nonparticipation continues even today as the US President is pulling the United States out of the 2015 Paris Agreement. Scientists warn that even if all signatory nations adhered to the agreement, it would still not be enough to protect Earth's inhabitants from some of the major effects of climate change mentioned earlier.

Even as the public becomes more aware of the global warming crisis, it is still business as usual for many. It is not that we don't care or that there is mass denial of the crisis. No, the majority of us believe global warming to be a threat to our existence and that it is caused by human activity. Often times our human nature is to ignore or minimize a problem until it hits us in the face. Consider how many people build a home in a known flood zone, near a volcano, on a hillside at risk of a mud slide, or close to the sea shore where the home is subject to major hurricanes and rising tides. Even as I am writing this, news of a volcanic eruption near New Zealand is being broadcast on the television. People were climbing all over the volcano, even though it had recently begun showing signs of becoming active, when it erupted resulting in the deaths of several people.

As the above example makes clear, we Homo sapiens are risk takers. Psychological studies reveal that risk takers either minimize the risk or maximize their abilities to deal with the risk at hand, or both. We see both behaviors in our response to global warming. While some are busy denying global warming and our own role in it, others are overly optimistic about our capacity to reverse global warming through geo-engineering, for example.

We can applaud and even marvel at the risks some are willing to take. Mountain climber Alex Honnold, for example, is held in high esteem by many for his courage. He knew full well of the risk to his very life when he climbed El Capitan without ropes. Clinging to the sides of the granite wall with nothing more than his fingers and the thin edges of rubber on his shoes, he completed the climb in slightly less than four hours when it would take days for climbers using ropes. He was just that confident in his abilities which he had tested many times before he began his historic climb up perhaps the world's most iconic cliff.

Like Alex, we take risks every day but, of course, not to the extent he did and does. Every time we leave our homes we are taking a risk. What we don't seem to realize, or perhaps do not want to realize, is that we are taking risks with our environment as we continue to rely on fossil fuels. The problem is that the risks we are taking with our use of fossil fuels are putting not only ourselves at risk but generations to come. Like the person who drives while intoxicated, we put others at risk with our actions. I believe, however, we are slowly becoming more aware of the risks associated with global warning and are beginning to act accordingly. More and more people are starting to take this crisis seriously. You must be one of them or you were not likely to pick up this book.

If climate change only affected those willing to take the risk, it would be acceptable to take that risk. But when our actions affect everyone, as they do on this spaceship we all jointly share, then it is an entirely different matter. I have deep respect for those willing to take risks. That is how progress is achieved. Risk taking has served the human race well over the two hundred or so thousand years of our existence; but, taking risks with Mother Nature is a whole other ball game! As is said, "Don't mess with Mother Nature."

Pioneering work into fear and risk by David Ropeik (2010) makes it clear that the closer in time an event is to us the more we fear it. Right now we all are afraid of the coronavirus because it is as close as the next affected person. The stranger we encounter at the grocery store could be a risk to our lives. We have less fear of global warming because, for most of us at least, it seems so far away. The level of our fear is not a good measure of the real risk. Global warming is arguably potentially worse than the coronavirus, but because it seems to be far off into the future, we fear it less.

We are at a crossroad in human history. With but one exception in the approximately 200,000 year history of Homo sapiens' existence has humanity ever faced a threat of this magnitude to our existence. During the cold war following WWII, humankind could have been totally obliterated with the unleashing of atomic weapons. Both the Soviet Union and the US had enough nuclear weapons to destroy humanity many times over, and practically all of life for that matter. Fortunately, cooler heads prevailed, sparing humankind a terrible fate like the one the Japanese faced during the Second World War. What we do or do not do now is of critical importance. Hanging in the balance is the health and wellbeing of all Earth's inhabitants.

We need to face the terrible fate that awaits humanity if we do not stop global warming. We have the opportunity to save this planet for future generations if we act boldly and quickly. If we fail to act accordingly, much of humanity, if not virtually all of it, will have to suffer from a climate that is not fit for human existence, at least as we now know it. Though the climate change prediction models cannot predict with one hundred percent accuracy what the future holds, if we continue to release carbon from the burning of fossil fuels into our atmosphere at the rate we are now doing, the models make it clear that humankind will be at risk for all kinds of dangers and disruptions.

We must stop listening to the doubters and take actions now. The earlier a potential disaster is faced the better the outcome of our actions. Evidence of not taking natural threats seriously enough can be seen not only in our response to global warming but also in our nation's response to the coronavirus. Initially, the President and even Dr. Fauci were not that concerned about it. Only when it reached near pandemic proportions did they became concerned enough to start mobilizing the nation to combat it. That lack of initial alarm, however, delayed our response to it and made the human and economic costs greater than they would have been had they started preparing earlier.

There is reason to be at least moderately hopeful. For example, most governments did not withdraw from the Paris Agreement. And some are even arguing that we need stricter standards. Consistent with our President's denial of climate change, he is withdrawing the United States from that agreement; but, the good news is all the other signatories are still committed to achieving its modest goals. In addition, many corporations have stepped up and are addressing the challenge in spite of the President's and his EPA director's foolish position. Some cities and states have taken up the cause and are enacting legislation to address the growing threat. Also, individuals, countless numbers of them, are beginning to take the problem seriously and make the necessary changes in their lifestyles to forestall global warming. Like gun violence, people are waking up to the fact that "thoughts and prayers" are simply not enough.

That great journalist, H.L. Mencken, while writing about the Scope's "monkey trial" in Tennessee, reached an alarming conclusion: "They know little of anything worth knowing, and there is not the slightest sign of a natural desire among them to increase their knowledge." Further, he went on to write: "…enlightenment, among mankind, is very narrowly dispersed." We have to believe that his remarks are specific to that time and to the issue of evolution being taught in public schools. If not, we have little chance of stopping global warming. Global warming requires an informed electorate without which effective changes in our energy policies will not happen.

Global warming/climate change is a war we can win if we are smart and if we care enough. We all need to be better informed, myself included. We can only make high quality decisions if we are sufficiently informed. All humanity deserves a healthy planet and a good life that can come from that. Generations to come are depending on us to protect our environment and leave them with an Earth that is friendly to life. We owe them that! And we need to act now.

Discussion Questions

1. Do you believe there is really a war going on? What are all the stakes in this war?

2. What does Amy Westervelt mean when she characterized the war as a "war of ideas?"

3. Do you know of anyone for whom the denial of anthropogenic (human-caused) climate change is an article of faith?

4. The Center for American Progress Action Fund identified four groups of those who deny climate change? Do you or anyone you know fit into one of those groups?

5. Do you see parallels between the Tobacco Wars of the 1950s and the war the fossil fuel industry is waging today? In each case, are the main arguments based on concerns about economics of the industry/investors or concerns about the environment/health?

6. Do you agree or disagree with Upton Sinclair's comments about the effect of money on the debate about global warming?

7. The author states that we are at a crossroads in human history. What does he mean and do you see it that way?

8. What do we really owe our children? Do we owe them a healthy climate? At any cost to us personally? Is it our problem or theirs?

9. Do you agree or disagree with Mencken's comments?

2

The Hard Facts

> *"...the emerging trend I'm most concerned about is purely human. The free world is lurching toward a polarized, post-truth reality that reminds me of my life in the Soviet Union, where the truth was whatever the regime said it was that day. If the battle for a shared, fact-based reality is not fought and won, 2030 will make the outrages and demagogy of 2019 look like a golden age of comity."*
>
> -Garry Kasparov, Russian chess master and writer

Gary Kasparov knows what it is like to live in a totalitarian state. In such a society the State controls the news. The State controls the "facts." One dare not think for him-or herself. One must follow the party line. Not to do so can lead to imprisonment or some other form of punishment. The state is the authority on all crucial matters. Kasparov sees democratic societies moving more and more in the direction of the repressive regime he left back in Russia. He warns of a battle being fought for the minds of people. He understands that it matters who wins this battle.

In his book (2000) Bishop John Selby Spong has a chapter titled, "The Heart Cannot Worship What The Mind Rejects: Entering A Dark Valley." This book chronicles his journey of faith from a young man full of beliefs shaped by his parents, his church and his Southern, conservative culture. He struggles to reconcile his faith with his emerging world view and biblical understanding.

I include it here because the title of that chapter captures the essence of the global warming issue. It is a truth our "hearts" do not want to accept. Many are struggling with the facts of global warming and are confused. Is it real? Is it a hoax? Is it a Chinese plot to take down capitalism? Is it a liberal foil to establish big government and a socialist state? What should I believe? What is the truth? What are the facts? Should I believe the science? Whom should I listen to, the President or scientists? The global warming issue is the new battleground for the minds of people. And since the fate of humanity is at stake, it matters gravely who wins this battle.

We, as a nation and as a world community, will not solve the problem of global warming as long as we remain confused or reject the scientific evidence of it. As long as we are willing to accept "alternative facts," whatever that is, our minds cannot discern fact from fiction, truth from falsehood. Until we are willing to enter the "dark valley of recovery," Earth will continue to warm and the climate will become more challenging and more dangerous. The recovery will be painful. One cannot accept global warming as a threat and engage in lessening its impact as long as one's mind rejects it. My hope is that this chapter will open the reader's mind to the reality of global warming and its primary cause. I hope to clear up much of the confusion and set one on a journey of discovery.

In this chapter a summary of the scientific facts in a Q&A format will be presented. The facts are undeniable that global warming is real and not a hoax and that it is caused by humans. The Buddha charges us to accept truth if after careful observation and analysis we determine it agrees with reason and if accepting it will benefit one and all. I believe the scientific facts meets those criteria.

Q. How do scientists collect data on global warming and is it reliable?

A. The search for information about our Earth is relentless and on-going. Data is collected by scientific instruments at thousands of places around the globe by US scientists, scientists from other countries, and from amateur weather watchers. Measurements are made from orbiting weather satellites, weather balloons, radar, ships at sea, weather buoys and thousands of land stations. NASA alone collects and analyses data from over 20,000 weather stations. Millions of pieces of data are collected each and every day. Data is collected from physical, chemical and biological material captured in ancient ice core samples, tree rings, coral reefs and layers of sedimentary rocks. Some of the data is over a million years old.

Science is not infallible. Scientists often debate the accuracy and meaning of their findings. This was especially true, for example, when Dr. James Hansen presented the conclusions from his research to a Congressional committee in 1988. His report created much controversy in the scientific community. Some thought he was being an extremist. Even today, for example, scientists do not agree on how hot the Earth will become or on the precise effects we can expect to see.

As far as reliability is concerned NASA reports that over two hundred, well-respected scientific organizations around the world agree with the following statement: "Observations throughout the world make it clear that climate change is occurring, and rigorous scientific research demonstrates that greenhouse gases emitted by human activities are the primary driver." NASA also reports that there is nearly 100 percent agreement among scientists on the facts of climate change and global warming.

Q. What is the difference between climate and weather?

A. Weather is what you get and climate is what you expect. Weather refers to local conditions around us in the short term -- hourly, daily, weekly or monthly. Climate refers to long-term averages of weather patterns over a longer period of time-over seasons, decades and years. Weather can vary from minute to minute sometimes, whereas climate is stable over longer periods of time. The fact that it may be hotter or colder on any given day, week or month does not disprove or prove anything about climate.

Q. What is the difference between global warming and climate change?

A. The terms are often used interchangeably as I will do throughout this book, but each has its distinct meaning. Global warming refers to the long-term heating of Earth's climate system over centuries of data collections beginning earnestly in the 1800s. Climate change refers to long-term changes in average weather patterns that reflect local, regional, and global climates. It is the warming of the Earth's surface and oceans that is causing changes in our climate. Global warming is the engine for climate change.

Q. How hot is the Earth becoming?

A. According to NASA's Goddard Institute for Space Studies (GISS), the average temperature of the Earth has raised by a little more than 2.0 degrees Fahrenheit (F) or 1.1-1.2 degrees Celsius (C) since the 1880s. Two-thirds of that warming has occurred since 1970! Temperatures are projected to rise another 2 to 4 degrees F over the next few decades. Why should such a small increase concern us? Even a two-degree change affects the Earth's climate. In the past a one to two-degree drop in Earth's global temperature was enough to usher in a little ice age. A five degree drop caused a large part of North America to be covered with a towering mass of ice 20,000 years ago.

A little rise in temperature across the land and sea surfaces of the Earth when averaged makes a big difference in our climate. Just as a little of habanera hot sauce on your chicken wings can make a huge difference, so does a small rise in the global surface temperature of the Earth. Who can tolerate even just a quarter teaspoon of wasabi on his or her sushi? And, just as some people's palates are sensitive to small amounts of certain condiments or spices, so the Earth's climate is very sensitive to small changes in temperature.

A recent report in the journal, *Reviews of Geophysics* (7/22/2020), reveals that if a doubling of CO_2 levels above preindustrial times occur, the Earth's atmosphere is likely to heat between 2.6 and 4.1 degrees Celsius instead of the once believed 1.5 and 4.5. Moreover, they predict that if current emissions continue, the doubling of CO_2 levels could happen well before the end of this century. "Climate sensitivity" is a disputed phenomenon among climate scientists. Some claim climate sensitivity is higher while others claim it is lower. Dr. Dessler, a climate scientist at Texas A&M University called the report "a real tour de force." He added, "It would be great if the skeptics were right. But it's pretty clear that the data don't support that contention." What this means is that it is more urgent than ever we curb our CO_2 and other GHG emissions.

Q. Has the Earth's climate ever changed before and should we be worried?

A. Yes, the Earth's temperature and atmosphere have fluctuated dramatically over the millenniums. Earth's early atmosphere consisted of large concentrations of poisonous gases such as hydrogen, methane, helium and carbon monoxide. Complex life could not have existed in such an atmosphere. Only after cyanobacteria, microorganisms related to bacteria but capable of photosynthesis, released sufficient oxygen into Earth's atmosphere could advanced life forms begin to exist. Cyanobacteria represent the earliest known form of life on the earth. Their release of oxygen not only made more complex life possible, it also provided life protection from harmful ultraviolet radiation from the sun by creating an ozone layer around the Earth to block the sun's harmful ultraviolet rays.

Earth has also warmed and cooled off many times over the course of 2.6 million years according to The National Oceanic and Atmospheric Administration (NOAA). The most recent glacier period occurred between 120,000 and 11,500 years ago. These fluctuations in Earth's temperature are caused by variations in Earth's orbit around the sun which is influenced by the gravitational tugs of the larger planets Saturn and Jupiter. However, according to scientists, the speed at which Earth is warming in the last two centuries is unprecedented in Earth's history and it is due almost entirely to human activity and not natural causes.

Scientists estimate that it takes about 5,000 years for the planet to warm between 4 and 7 degrees C. The warming of the past century, 0.7 degrees C, is roughly eight times faster than ice age- recovery on the average. Paleoclimate data show that atmospheric CO_2 levels are higher than they have been in the past 800,000 years. Yes, we should be worried because the changes in climate we are witnessing now will last, are potentially disastrous and will not be easily reversed.

NASA models predict the Earth will warm between two and six degrees Celsius over the next century. When global warming has occurred at various times in the past two million years, it has taken Earth about 5,000 years to warm five degrees. NASA's predicted rate of warming for the next century will be at least 20 times faster! This is due principally to the increased concentration of CO_2 in the atmosphere due to human activity.

Q. What is causing the Earth to warm?

A. Earth's atmosphere serves as a blanket over the Earth keeping it at the right temperature. Our atmosphere now consists mostly of nitrogen (78%), oxygen (21%), and a collection of trace gases (carbon dioxide, water vapor, ozone, argon, methane). Even though CO_2 makes up only a small portion of our atmosphere, 0.04%, CO_2 operationally is the most critical gas as far as Earth's temperature is concerned. Too much CO_2 and Earth is too hot. Too little CO_2 and we would be freezing. Any change in the amount of CO_2 causes a change in Earth's temperature since CO_2 is a powerful regulator of Earth's temperature. Changes in the concentration of other gases, such as water vapor and methane, for example, also affect Earth's temperature, but CO_2 is the primary driver of climate change. Think of CO_2 as a thermostat that controls the temperature of the Earth.

Methane is also becoming a very important factor in global warming as more and more of it is being released from oil well production sites. In America alone methane is being released at twice the rate previously estimated based on new satellite data. Scientists estimate that, pound for pound, methane can warm the planet greater than 80 times as much as CO_2 over a twenty year period. The chemical composition of our atmosphere is critically important. We mess with it at our own peril.

Here's how our atmosphere affects our temperature. Atmospheric gases both absorb and reflect the sun's radiation back to Earth, especially infrared radiation, once it is reflected off the land and sea surfaces of the Earth. These gases serve as a blanket over the Earth. The more CO_2 in the atmosphere the thicker the blanket, so to speak, and more of the sun's energy is kept inside this blanket. The more energy from the sun trapped inside our atmosphere, the hotter Earth becomes, which, of course, is what we are witnessing today.

Q. Where is the increase in CO_2 coming from?

A. CO_2 is critical for life on this planet. But too much is not beneficial to life in Earth. Data from the early nineteenth century on, since the Industrial Revolution, has revealed a steady increase in CO_2 in our atmosphere, especially in the last few decades. The burning of fossil fuels (coal, gas and oil) releases CO_2 as a byproduct which increased dramatically beginning with the Industrial Revolution. Fuels were needed to run the machinery of commerce that was providing the goods that people demanded as well as the heating and cooling of our homes. Since the Industrial Revolution our use of fossil fuels in industry skyrocketed as our population was also increasing. The result is more people using fossil fuels as well as exhaling CO_2 into the air. In addition, to a lesser degree, deforestation due to the burning of large areas and the cutting of trees, such as in the Amazon, to make room for crop or the grazing of cattle is adding to the problem.

The result is a buildup of CO_2 in our atmosphere. For example, ice core data shows that average ice age CO_2 concentrations were 185 pp. and pre-industrial was 278 ppm. The data is conclusive, global warming is being caused by human activity, namely, the release of CO_2 into our atmosphere. The NASA graph below shows how rapidly CO_2 levels have risen in the last few years compared to 800,000 years ago. CO_2 levels now exceed 400 ppm and are continuing to rise.

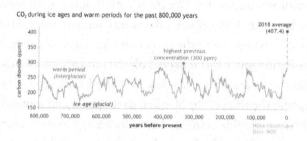

Q. Couldn't the global warming we are seeing be caused by other factors?

A. No! There is much evidence for anthropogenic global warming. The increase in green house gases (GHGs) since the 1880s parallels the increase in Earth's temperature. In addition, there is no evidence for other causes. The warming of the Earth is happening at a pace much faster than ever recorded except following a major and sudden event on Earth such as a strike by a very large meteor or the eruption of a major volcano. Whereas there have been warmer periods and colder periods over the course of Earth's history, the warming is happening faster now than it has in the past million years, roughly eight times faster than the average ice age warming recovery.

Scientific data shows that changes in the sun's output, major volcanic eruptions, cycles such as El Nino and the Pacific Decadal Oscillations, and oscillations in the Earth's orbit around the sun cannot fully account for the recently rapid rise in Earth's temperature. Paleoclimate data shows that atmospheric CO_2 levels are now higher than they have been in the last 800,000 years. There is some evidence that indicates that it may be more like over two million years since there has been such a high concentration of CO_2 in our atmosphere.

The NASA graph below shows the effect of human activity on changes in the Earth's temperature. This data, along with mountains of other data, tells scientists that the current warming we are experiencing could not be caused by natural factors.

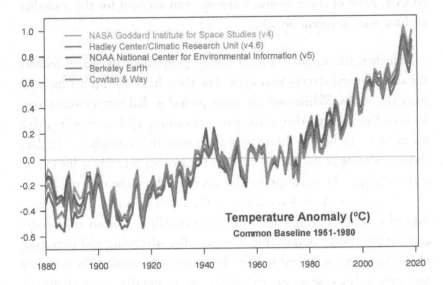

NASA has concluded: "There is no plausible explanation for why such high levels of CO_2 could not cause the planet to warm." And, further, NASA reported: "The current warming trend is of particular significance because most of it is *extremely* (italics mine) likely (greater than 95% probability) to be the result of human activity since the mid-20th century and is proceeding at a rate that is unprecedented over decades to millennia."

Q. But, is it not true that natural processes can cause increases and decreases in global temperatures?

A. Yes, but.... Key feedback systems do cause *minor* changes in local or regional temperatures but not to the scale we are now witnessing globally. For example, global warming does increase the amount of water vapor in the atmosphere which causes an increase in temperature, but these changes do not account for the increases we have seen globally since the Industrial Revolution. In addition, decreases in snow and ice cover causes warming because less sun radiation is reflected back into space. The effects of clouds have not yet been determined. According to NASA and NOAA, none of these feedback systems can account for the warming we have seen in recent decades.

Further, the amount of solar energy received by the Earth follows the sun's natural eleven- year cycle, but there has been no net increase since the 1950s. While over the same period global temperatures have increased markedly. Also, if the sun was causing global warming, this warming would be seen throughout all layers of the atmosphere. The data shows warming at the Earth's surface, the troposphere where life exists, and cooling in the stratosphere, the layer just above the troposphere.

The sun does have some effect on the Earth's climate. That effect is limited to subtle changes in the Earth's axial tilt as it orbits around the sun not the energy output from the sun. The advancing and retreating of the ice ages are linked to the tilt of our orbit around the sun, but temperature increase seen in recent decades are not due to any change in Earth's orbital pattern since there has not been any significant change in that pattern in the last two hundred years when the Earth's temperature began to rise significantly.

Q. What is the evidence that climate change is real?

A. Listed below are many indicators that climate change is happening:

1. Global temperature rise: There has been a rise of 2 degrees F, 1.1 degree C, since the late nineteenth century. Two-thirds of this warming has occurred since 1970. As stated earlier, temperatures are projected to rise by another 2 to 7 degrees over the next few decades.
2. Warming oceans: Oceans cover two-thirds of Earth's surface. The average temperature of the oceans has increased more than 0.4 degrees F since 1969. Oceans retain heat longer than land as anyone living near a large body of water knows. This becomes

important as we shall soon see when reversing or slowing global warming is a concern.

3. Decreasing ice sheets: Greenland ice sheets have declined in mass some 286 billion tons per year between 1993 and 2016 and continues to decrease each day. The Antarctic ice sheet lost 127 billion tons during the same time period. The rate of Antarctic ice mass loss has tripled in the last decade.

4. Glacier retreat: Almost everywhere in the world glaciers are retreating. This includes the Alps, Himalayas, Rockies, Alaska and Africa. All this is measured by satellites. The US's own Glacier National Park will be without glaciers if this trend continues. Since 1961, the Earth has lost the equivalent of a block of ice as large as the continental United Sates at a depth of 16 feet thick!

5. Decreased snow cover: Satellite data reveals the amount of snow cover in the Northern Hemisphere has declined over the last five decades and that snows are melting sooner.

6. Sea level rise: Global sea levels have risen by about eight inches in the last century. The last two decades of the last century saw a nearly doubling of that of the previous century and is accelerating each year. Part of the reason for this is not just that glaciers are melting, but also because heat causes objects, in this case water, to expand.

7. Arctic sea ice: The last several decades has seen a diminishing of the extent and thickness of the Arctic sea ice. It is predicted in that in a few years the Arctic sea will be navigable in the summer months which have never happened before.

8. Extreme weather events: Record high temperature events have been increasing all over the Earth. Almost every continent has reported unbearable, unforeseen high temperatures. For example, a report by the *World Weather Attribution*, (7/15/2020), concluded that the prolonged heat wave in Siberia this summer (a record breaking 38 degrees C, 100.4 degrees F) in Verkhoyansk on June 20) was made "at least 600 times more likely as a result of human-induced climate change." In addition, higher

numbers of extreme rainfalls, tornados and hurricanes have been reported around the globe. Prolonged droughts have been more frequent too.

According to the *World Meteorological Organization*, every storm has the same DNA-- moisture, unstable air and something to ignite the two upwards which is often heat. When the Earth warms, hot wet air rushes upwards in huge columns where it collides with cool dry air. This forms volatile cumulus clouds that can swell at the top of the troposphere. Depending on the temperature of the air, as much as a million tons of water can be lifted up into the atmosphere. The hotter the air, all else being equal, the more moisture there is to fuel the storm. The more moisture to fuel the storm, the more likely a violent storm will be the result. Budding storm cells are formed and when one of these columns punches through the tropopause, the boundary between the troposphere and stratosphere. The result is the storm grows larger, growing on the energy-rich air of the upper atmosphere. As it continues to grow the violence of the storm, be it rain or hail, increases.

As it turns out the perfect storm-birthing region in the U.S. is the Great Plains. Many of the world's most dangerous storms are formed here. In the spring and summer months, moist air from the Gulf of Mexico mixes with dry air from the Arctic and southwestern deserts. This volatile mixture is corralled by the Rocky Mountains. This combination of events often leads to the outbreak of many violent storms.

9. Ocean acidification: Since the beginning of the Industrial Revolution, ocean acidity has increased by about 30%. The oceans absorb CO_2 which combines with H_2O, water, to produce carbonic acid which is harmful to fish, microorganisms and coral life. This acidification is threatening all of our ocean food sources and disrupting life and food cycles for all marine animals and organisms.
10. Permafrost: Permafrost is land that has been mostly frozen, some for half a million years or more. About a fifth of the Northern Hemisphere landmass is permafrost. It is thawing at

unprecedented rates. As it thaws bacterial decomposition of the organic matter in the permafrost occurs releasing huge volumes of methane gas, a very potent GHG into the atmosphere, thus supercharging global warming.

Q. Ok, but how does all this affect me?

A. Let's start with the fact that our climate is changing mostly for the worst. Over the last 50 years we have seen dramatic increases in prolonged periods of excessive heat, heavy rains, and in some regions, severe floods and droughts. For example, the drought that affected parts of Texas and Kansas in 2011 caused $10 billion in crop loss alone which meant we all paid more for our food. Floods cause untold hardship and damage. People living along rivers are seeing bigger floods than ever has been seen in recent history. Floods cause great hardship as well as property loss. People living near oceans are experiencing flooding they have never experienced before, especially during high tides or hurricanes storm surges compounded by increased rainfall. Rising ocean levels are polluting fresh ground water with salty ocean brine making it not fit to drink. This is especially a concern in southern Florida, where rising sea levels threaten not only people's property but also their drinking water. Also, folks living along the coasts are worried about the future of their homes and properties.

Heavy precipitation has been seen in every region of the U.S. between 1958 and 2012. According to NOAA, The National Oceanic and Atmospheric Administration, the greatest rise has been in the Northeastern states (71%) while the least in the continental U.S. was the Southwest (5%). In 2018 there was a March nor'easter that caught everyone by surprise, a frigid and snowy April from the Northern Plains to the Upper Midwest, and great coastal flooding accompanying Hurricanes Florence and Michael. Data indicates there has been a substantial increase in all measures of Atlantic hurricane activity (strength, duration and increased rainfall due to ocean warming) since 1980. Increased temperatures affect the most vulnerable among us, the poor, children and the elderly. Melting permafrost creates unstable soil which causes buildings to buckle and collapse.

Air pollution is a serious problem for many with breathing disorders such as asthma and COPD, for example. The Lancet Commission on Pollution and Health (10/19/2017) estimate that pollution is responsible for an estimated nine million premature deaths in 2015 (16% of all deaths worldwide--three times more than from AIDS, tuberculosis and malaria combined) which would not likely have occurred if it were not for pollution. Not only is climate change affecting humans, it is affecting animal life as well. Changes are already occurring in animal natural habitats forcing animals to migrate north to avoid the rising temperatures. Insects that were heretofore confined to southern climates will invade the northern regions. This will bring tropical diseases that haven't normally been exposed to as well.

These are but a few of the negative impact climate change is having on our lives. And, if the scientific climate change models are right, and there is no serious reason to doubt the many models predicting similar outcomes, the worst is yet to come. It cannot be said for certain that a singular extreme weather event is caused by climate change, but one thing is certain, extreme weather events become more probable as the climate becomes warmer. Extreme weather events are more common now than at any point in human history. The National Climate Assessment data makes clear, "Human-induced climate change has already increased the number and strength of some of these extreme weather events." And, "Evidence indicates that the human influence on climate has already doubled the probability of extreme heat events…"

All scientific models point to a future that is challenging to say the least. Major disruptions in our everyday lives will occur as noted above. Predictions have value even if no one knows with 100% accuracy what the future will bring. Noted author, Octavia Butler addressed the value of predictions in her essay (2000), *"A Few Rules for Trying to Predict the Future."* Written before climate change modeling was as accurate as it is now, she wrote: "So why try to predict the future at all if it's so difficult, so nearly impossible? Because making predictions is one way to give warning when we see ourselves drifting in dangerous directions. Because prediction is a useful way of pointing out safer, wiser courses. Because, most of all, our tomorrow is the child of our today. Through thought and deed, we exert a great deal of influence over this child, even though we can't control it absolutely. Best to think about it, though. Best to try to shape it into something good. Best to do that for any child." To which I say, Amen!

Q. Can global warming and climate changes be stopped?

A. Yes, but..... Global warming cannot be stopped right now even if we stopped burning fossil fuels altogether. Unlike the coronavirous, global warming does not peak. It would only "peak" if we stopped burning fossil fuels. But, even if we stopped polluting our environment today with GHGs, it would take several generations for CO_2 levels to return to a baseline of 350 ppm that most scientists say is more ideal for human life. It is now at 421 ppm and increasing. This delayed reaction in cooling is called the "thermal inertia" and is due in large part to two factors. Much of this heat is stored in the oceans which covers two thirds of the Earth's surface. Oceans release their heat much slower than land or air. This explains, for example, why temperatures in Cleveland, Ohio, which is on Lake Erie, are, all things being equal, often warmer or cooler than in Mansfield which is further away from the lake. Also, CO_2 remains in the atmosphere for hundreds of years. Even if we stopped burning fossil fuels altogether today, scientists estimate that the surface temperature would still increase a little more than 1 degrees F. The natural processes that absorb CO_2, oceans and photosynthesis, simply cannot absorb enough CO_2 in a short time.

Stopping global warming is a monumental task, bigger than any task humanity has ever faced I would argue. Global warming is a problem that is worldwide in scope, affecting every life form on this planet and whose effects will be with us for generations. I will return to this theme again in Chapter Eight.

Q. What is meant by the term "carbon footprint?"

A. Carbon footprint refers to the amount of CO_2 a person is responsible for emitting into theatmosphere through one's daily activities. These activities include such activities as driving a car or riding a bus or train, heating and cooling one's living space, the foods eaten, consuming goods of all kinds and even breathing adds to one's carbon footprint. Literally, everything one does leaves a carbon footprint. The average U.S. citizen daily footprint is approximately 115 pounds of CO_2 or 21 tons per year. This is four times the global average since we consume more food and manufactured goods and have a higher standard of living than most of the world. One's carbon footprint can be calculated easily and there are many websites one can use to do so.

Q. Is there any hurry to solve the problem of climate change?

A. Remember once Humpty Dumpty fell, he could not be put back together again. Scientists warn us that once the concentration of CO_2 reaches a certain level, it will take generations for it to return to a more livable level. Projections of future climate conditions are based on results from climate change models. These models, based on complex computer programs, predict how the climate is expected to change under different possible scenarios called Representative Concentration Pathways or RCPs. The reader is referred to Graham P. Wayne (2013).

The picture does not look good, but it's also not clear. According to one model, for example, we could see increases in the range of 5.2 degrees F. This would have disastrous results on all life. More optimistic projections do exist, but it is a matter of how soon and how robust our efforts are to reduce the emission of GHGs. Time is of the essence if we are to avoid catastrophic effects. Robust efforts must replace fantasies about how we can control the amount of CO_2 in our atmosphere. Planting a trillion trees as proposed by the World Economic Forum recently, is a long shot, to say the least. Even if we were to accomplish such a feat, which is highly doubtful, it would be at least ten years in a best-case scenario before the trees would have any measureable effect on the CO_2 concentration in our atmosphere. We simply cannot wait that long before we start reducing our carbon footprint.

It should be noted that *actual* future GHG concentrations depend on what we do now to control our emissions of GHGs. If we continue to emit GHGs at the level we are now or if we decrease our emissions and do so quickly, the results will be different. The effects of climate change will be better or worse depending on what we do now about our GHG emissions.

No, if we are to lessen CO_2 emissions and reduce global warming, we must make realistic and robust efforts to do so. Commenting on the record temperature of 64.9 degrees F recorded in Antarctica on February 6, 2020, one researcher said: "This is the foreshadowing of what is to come. It's exactly in line of what we've been seeing for decades." So, yes we must act now and act with realism, not fantasy or fancy. The time for dodging the issue has long passed.

Now if these facts are new to you or if they contradict with what you already believe, perhaps it is a matter of source. Source matters. There is no lack of "facts" about climate change just as there was no lack of "facts" about the effects of tobacco on human health during the Tobacco Wars of the 1950s. The tobacco industry was initially successful in confusing the public and legislators about the effects, but soon the science caught up with their misinformation campaign. Now the fossil fuel industry and its followers are having at least some success in confusing the public and delaying effective action.

Unlike the 1950s, however, there is a more pernicious effect in play making it even harder for people to know what to believe. Social media has changed the debate. It not only changed the way information or news is received but also the reliability of it. We used to be able to trust the source. If Dan Rather, Walter Cronkite, or Tom Brokaw said it, you could take it to the bank. Or, if it appeared in a major newspaper, it was most always true and could be trusted. Editors existed to separate fact from fiction and opinions from facts.

Kenneth Miller (2020) in an essay titled, "Your Brain on Tech," wrote: "While the upsides of all this pixel-gazing are plentiful, the downsides can be scary. In the public arena, online filters generate bubblewt (sic), reinforce our preconceptions and amplify our anger. Brandishing tweets like pitchforks, we're swept into virtual mobs; some of us move on to violence IRL. Our digitally enhanced tribalism upends political norms and sways elections." Social media can be a hostile place and bad for the climate.

In addition, today anyone with access to social media is an "expert," a "reliable" source of information. There are no editors reviewing the news on social media. I recall during the presidential campaign of 2016 someone reported that Hillary Clinton was running a child sex ring out of a certain pizza place in Washington, D.C. A man went there with a gun to put a stop to what he believed was a child sex ring. Fortunately, a disaster was avoided, but this illustrates the power of social media. There are so many more examples that could be given of the dangerous influence of social media on global warming as well other issues.

Far too many people depend solely on social media for their information and facts. It is a virtual lawless jungle of information out there on social media. It serves many useful purposes to be sure, but it is like any tool; one must be careful how it is used. Today, it has become an effective tool for those who want to confuse the public about global warming. In fact, there is an "expert" on social media for any belief one wants to hold. "Flat Earthers," for example, have their "experts" touting the facts about the Earth being flat. So many other groups have their "facts" too. One can readily find support on social media for any belief.

With a matter as critical as global warming and climate change, we must get our facts straight. Let me be clear. The fact that our Earth is warming and that such warming is due to human activity is not debatable. What is debatable, for example, is how hot Earth will become, what the long-term effects of global warming will be and how and when we should act to counter it. There is no room for error. The fate of humanity depends on facts not beliefs or alternative facts.

The sources of the facts just presented are the scientists themselves reporting in their peer-reviewed journals, professional associations, and private and public institutions interested in the facts about climate change. If these facts don't agree with your "facts," then check your sources. The source is as important as the fact it presents! If you are having trouble accepting these facts, then read the following chapters which addresses the difference between the reliability of beliefs and the psychology of disbelief or denial.

One final thought. The International Panel on Climate Change (IPCC) concluded (2019): "Taken as a whole, the range of published evidence indicates that the damage costs of climate change are likely to be significant and to increase over time." Every person on this planet will be negatively impacted one way or another by climate change sooner or later. Only those living in more affluent countries with more resources will be affected less. Individuals with greater resources, the richer among us, will fare better than the poorer who are more vulnerable. Climate change is or soon will be everyone's problem. No one will escape the negative impact of climate change.

All should understand that climate change, if left unabated, will not be a mere inconvenience; it will be a disaster especially among the poorest people and nations. Climate change will create a disaster like we have never seen before.

Readers interested in more data are encouraged to visit the websites of NASA and NOAA. There are other reliable sources as well which are listed at the end of this book. Just be sure that whatever source you turn to is a credible source backed by peer-reviewed scientific data as there is much misinformation being spread by organizations with scientific-sounding names.

Discussion Questions

1. Is what Gary Kasparov wrote true only of Russia? The US?

2. Is your "heart" (reference Bishop Spong) having trouble accepting the facts of global warming?

3. Are you inclined to accept the science of climate change or doubt it?

4. Are there any facts the author presents that you would take issue with or don't quite understand?

5. Of the ten indicators of climate change, have you personally experienced any or know of someone who has or is?

6. Do you know what your personal carbon footprint is or have any interest in finding out?

7. Do you see the poor suffering from climate change more than the more advantaged as we have seen (saw) in the coronavirus pandemic? If so, why is this so?

3

Not All Beliefs Are Equal

"Everyone is entitled to his own opinion, but not his own facts."

-Senator Daniel Patrick Moynihan

"The dictators of the world say that if you tell a lie often enough, people will believe it. Well, if you tell the truth often enough, they'll believe it and go along with you."

President Harry Truman

Beliefs can be dangerous to one's health and wellbeing. A friend of mine, for example, once discovered a lump on his back. He believed it was nothing serious and, therefore, did not go see a doctor. When it continued to grow and started to become painful, he became concerned and made an appointment to see a doctor. By then, however, it was too late. The cancerous tumor had metastasized and had reached the point where treatment would not help. Unfortunately, he had waited too long. His denial and neglect cost him his life.

Likewise, what we believe about global warming is of the utmost importance. Fundamentally, global warming is not a matter of belief. No one ever asks if you believe in the moon, for example. We all accept the moon is real because we can see it. Well, scientists are telling us that global warming is as real as the moon. The evidence is undeniable that global warming is a real phenomenon and that it is caused by human activity, especially the burning of fossil fuels. Global warming can be "seen" in the vast volume of data that has been collected by scientists since the late nineteenth century.

People do not always welcome facts, however. My mother used to say, "If there's ever anything seriously wrong with me, I don't want to know." Many have said the same. It appears some are saying that about global warming. If it's real, I don't want to know. Facts can be disturbing, especially if one believes there is nothing that can be done to correct the problem or improve the situation. For many it seems that it is better to remain ignorant about a problem and not worry about it than to know there is a problem and not be able to do anything about it. Global warming, however, is a problem we can do something about.

Too large a number, 16%, of Americans continue to deny global warming. Even slightly over a third, 36%, report it is not personally important to them. There are also some who believe the problem is being exaggerated. Such beliefs are delaying our response to global warming, which is only making global warming more of a threat to our health and wellbeing. Even though there is something seriously wrong with our planet, many apparently would rather remain in the dark about global warming than actually face the reality of it. Denial and ignorance do not lead to sound policies.

False beliefs, beliefs not based on facts, about global warming are many and varied. For example, it is the belief of one man I know that global warming is a natural phenomenon and is nothing new; therefore, he insists, it is not our fault and we cannot do anything about it. He claims the Earth has been steadily warming since the last ice age and that is part of God's plan. To which he quickly adds, God is still in charge and knows what "He" is doing. Such people believe that whatever happens is God's will.

This belief is his truth and, of course, provides him with the comfort of a simple answer to a complex problem. We too often prefer simple answers that are comforting to complex answers that are disturbing.

H.L. Mencken famously said, "For every complex problem there is an answer that is clear, simple, and wrong." Denying or ignoring global warming is an example of what Mencken was speaking about. Complex answers create cognitive dissonance, a topic that will be explored in the next chapter.

This man clings to this belief in spite of the fact that there has been a dramatic increase in average global temperature since the Industrial Revolution, especially in recent decades, due to the buildup of CO_2 in our atmosphere. Such a dramatic increase due to the gradual increase of CO_2 in our atmosphere is unparalleled in the history of our planet. This increase cannot be explained by naturally occurring phenomena such as increased solar energy, for example. He can also think God is in charge, but that doesn't change the fact that many are already suffering from global warming and more will as the problem grows worse.

His is not the only myth confusing the issue and delaying robust action on climate change. Here are just a few others:

- The Earth's climate has always changed and it is no different now.
- Plants can absorb enough CO_2 evolving to higher levels of consciousness or spirituality to prevent serious global warming.
- Global warming is not real because it is still plenty cold in many parts of the world.
- Climate change is a future problem, not one that we need to solve now.
- Increased solar energy from the sun is the cause of global warming.
- The interest in renewable energy is nothing but a money-making scheme.

Myths are beliefs not based on facts. Such beliefs are not grounded in science. Beliefs not based on the facts of science have two obvious limitations. These limitations are especially true of strongly held beliefs. The first limitation is that beliefs often masquerade as truths. Because beliefs can *feel* so right, many mistake them for the truth and cling to them as if they are the truth in spite of evidence to the contrary.

Many who find the *facts* about global warming disturbing adopt a *belief* to counter it. They convince themselves, or allow themselves to be convinced, that their belief is a "fact." A false belief is an intellectual dodge, a way of justifying a position and avoiding an "inconvenient truth" as Al Gore would say. A false belief is also a lazy belief. A belief based on facts is often not easy or quick to discover and may not feel good, but facts are the only solid basis for making decisions or policies.

The second limitation is related to the first, but it presents the greatest danger of strongly held false beliefs. Such beliefs prevent inquiry. They kill curiosity. When we assume our belief is true, we quit searching for other possible answers. At that point our minds are closed. We no longer are open to new information, especially information that would challenge our belief. It's the "my mind is made up; don't confuse me with the facts" phenomenon.

We all should be concerned about adopting and clinging to beliefs not based on facts. Significant progress will not be made on global warming until facts-- not false beliefs or myths-- influence our response to it. Currently, such beliefs are driving the debate about global warming and what to do about it. Our efforts as a nation and a world community to address climate change are being hampered by people holding onto beliefs not based on facts.

This raises the question of who and what to believe. When it comes to such critical issues such as those involving our own health or the health of our planet, we must rely on science. You must understand there is an "information war" going on in our society that pits scientific facts against flawed beliefs. As Shoshana Zuboff writes in her book (2019), it is a war about: "Who knows? Who decides who knows? Who decides who decides who knows?" Within reason, we should trust the scientists not politicians.

If science does not decide, we and generations that follow us are forever doomed to pay the price of our folly. We discovered how important it was during the coronavirus pandemic to make decisions based on science, the data, not someone's gut feeling or opinion. Scientists, not politicians or religious leaders, must be the arbitrators of the truth about global warming. If we fail to embrace the science of global warming, we are doomed to make poor decisions regarding this crisis, decisions that will negatively impact many generations to come.

The scientific method is a trusted method for determining fact from fiction. We use science in medicine every day. We trust science when it comes to matters of our health, in matters of life and death. The scientific method requires repeated examinations of facts and testing and verification of data. Data that meets strict scientific criteria will be published in peer-reviewed journals not in magazines or non-peer-reviewed journals that may have scientific sounding names.

Science is not a partisan endeavor. Its data is not subjective. It must be noted, however, that scientists don't always agree. Fierce debates sometimes take place in scientific circles. For example, the science was not settled on anthropogenic global warming when Dr. James Hansen testified to Congress in 1988. Data that is consistent and studies that can be replicated form the basis of solid scientific conclusions. Science is not perfect, but the scientific method has proven itself invaluable. The reader who wants more scientific data on global warming can start with trusted sources such as NASA or NOAA or professional journals.

President Harry Truman (*Where the Bucks Stops*) understood the power of truth and fiction. Disciples are won to a point of view not only when it suits their purposes, but when they keep hearing the same thing over and over again. It doesn't matter if it is the truth or fiction, repetition serves the function of reinforcing the message.

Source of information about climate change must be sources that have proven to be reliable, sources based not on opinions, but on facts. The future habitability of our Earth and the quality of life of generations to come depend on us making decisions based on facts not fiction. For your own sake and the sake of future generations do not settle for unproven beliefs. If your beliefs are not supported by science, abandon them and seek the facts. While it may be true that beauty lies in the eyes of the beholder, facts do not.

Discussion Questions

1. Do you agree or disagree with the author that anthropogenic global warming is not a matter of belief but one of fact? What distinction between facts and beliefs is he making?

2. Do you believe the issue of global warming/climate change is being exaggerated? If so, for what gain?

3. Do you hold any what scientists call myths about global warming or know of someone who does?

4. The author writes that a false belief is a lazy belief. Do you agree?

5. How does Zuboff's statement, "Who knows? Who decides who knows? Who decides who decides who knows?" apply to global warming? In other words, does source matter?

6. Can you think of any information or opinion that became accepted as factual because it was repeated many times?

7. What is the source of your information about climate change? Do you rely on just one source or seek information from a variety of sources?

4

Why Facts Don't Matter

"A man with a conviction is a hard man to change. Tell him you disagree and he turns away. Show him facts or figures and he questions your sources. Appeal to logic and he fails to see your point."

-Leon Festinger, Social psychologist

My hope is that this chapter will help you become a connoisseur of facts. Our personal growth and wellbeing as well as the health of this planet require that we become experts on ourselves and global warming. Approaching the subject of ourselves and global warming with an open mind and with a hunger for the truth is of utmost importance. For progress to be made on global warming all of us must become more aware of how and why we hold the beliefs we do.

If after reading this chapter you find yourself having a better understanding of why you believe the way you do, a more open mind to views that may make you uncomfortable, a greater willingness to give serious thought to ideas that perhaps earlier you would have rejected and more willing to embrace beliefs that might put you at odds with your group, then you will be on your way to accepting the scientific facts of global warming and, thus, becoming a better friend of this planet.

As noted in the previous chapter, humans have a tendency, even a need, to dismiss facts that threaten their emotional equilibrium. Facts that run counter to our cherished beliefs create emotional turbulence; hence, they are rejected in favor of a belief that is more emotionally acceptable. For our peace of mind we need to maintain a state of emotional equilibrium or balance.

For that and other reasons which we will soon examine, facts have not always fared well in human history. For example, when Galileo challenged the Church's dogma that the Earth was the center of the universe (the known universe at that time, of course), the geocentric model of the universe, he was punished by the church. Galileo was simply verifying Copernicus's heliocentric model of the universe that recognized the sun at the center.

Galileo was required to "abjure, curse, and detest" his position and was sentenced to house arrest for the remainder of his life. It must be noted that Galileo's "house" was actually his villa in Tuscany and that he had the service of a maid, his daughter who was a nun. In addition, he continued to publish scientific papers.

Galileo's case illustrates the fact that those who hold authority are reluctant to relinquish it. Authority never welcomes challenges. We are witnessing that today in the global warming debate where the government and even some elements of religion are challenging the facts or authority of science. This can also be seen in the debate about when life begins. Both issues are seen as the province of God. The worry seems to be if science is recognized as the authority on matters associated with the faith of some believers, then other matters of faith will be questioned and perhaps abandoned too by the faithful.

One other example will be illustrative. Facts were also attacked during the 1950s by the tobacco industry. As more and more scientific evidence became available proving that tobacco was harmful to human health, the tobacco industry went to war (now known as the Tobacco Wars) against the facts. Instead of directly disputing the facts, which was a battle they would thought they would lose, they published their own "facts" from their own scientists which confused the public and lawmakers. The result was a delay in passing legislation to protect the public. Even though this approach was initially successful, as more and more evidence mounted, Congress finally acted allowing the Surgeon General to declare tobacco a serious danger to our health.

Today, as presented in Chapter 1, the fossil fuel industry has taken a page from the play book of the tobacco industry. It has its scientists denying anthropogenic, human-caused, global warming even though they know otherwise. (Interestingly enough, they sometimes even admit to anthropogenic global warming, which only adds more confusion to the debate.) For example, Exxon Mobil ads touting biofuels, such as algae, is a tacit admission that global warming is real and is caused by the burning of fossil fuels.

As noted in Chapter 1, the fossil fuel industry knew in 1982 that the burning of carbon-based fuels was causing a rise in the Earth's surface temperature. They also knew that the problem would only get worse as the internal Exxon memo mentioned in that chapter makes clear. But, because of vested interests, the fossil fuel industry never alerted the public. The industry felt no responsibility to do so as their loyalty was to their shareholders.

These attempts to confuse the public or keep us in the dark by posing to be an authority remind me of a story, *The Limitations of Dogma*. In the story the great Sultan Mahmud saw a porter dragging a heavy stone through the streets of the city. Having compassion for the man, the Sultan ordered him to leave the stone where it lay which he was more than relieved to do. The problem was that the stone became an obstacle to all who had to pass by. Requests to Mahmud to have the stone moved were not granted as Mahmud did not want people to think his orders were just whims and not based on sound judgment. So, the stone laid there forever out of respect for royal commands.

People reacted differently to the royal command. Some though it was evidence of how stupid it was of authority to try to maintain itself even when it was obviously foolish to do so. Others believed it important to respect all royal commands regardless how whimsical or inconvenient they may turn out to be because royal commands represent authority. A third group thought the order should serve as an example of how those who rule by inflexible dogma should not be in positions of authority or power.

Today we see the same scenario playing out as we debate what to believe and do about global warming. The question really, is whether or not we will move the proverbial stone. Will we address the problem of global warming with common sense, objectivity and the facts of science? Will our concern for the general welfare of the planet and its people rise above our self interests? Or, will adherence to dogma, conformity to authority, and economic, political and personal interests triumph over scientific facts? The verdict is out, but we have to hope, for the sake of the planet, that scientific facts will win the day.

One can find any number of "experts" and publications today disputing the science that global warming is real and caused by human activity. If you read any of the propaganda, read it with a healthy degree of skepticism. The tactic is to take a truth or a half truth and spin it to support a false argument. For example, I once heard of a man who declared that since the atmosphere of the Earth has changed many times over its 4.5 billion year history (true), the changes being seen now are just part of the natural atmospheric cycles of the Earth (false). These so-called experts are practicing what I call "junk science," drawing false conclusions from insufficient or incomplete data.

Thus far they have been successful in confusing the public and delaying robust action on global warming. To effectively challenge the prophets of lies and misinformation, their motives and tactics must be understood. We will discover that the motives to justify the rejection of scientific facts are many. In addition, those who deny also use clever psychological mechanisms for dismissing the troubling facts about global warming. We will examine both of those dynamics, motives and psychological mechanisms.

People cling to misinformed opinions and downright falsehoods about climate change for a variety of reasons. One motive for denial is to avoid what is called cognitive dissonance. Cognitive dissonance occurs when one's beliefs or values clash with evidence or when one's behaviors are in opposition to one's values. Stress is created when people try to hold two or more opposing views or when people act against their own principles. For example, one cannot believe in anthropogenic global warming and at the same go out and buy a gas-guzzling Hummer without perhaps creating some emotional discomfort such as guilt or some degree of self-contempt.

One way people use to avoid cognitive dissonance is just to deny global warming or to believe it is not the threat scientists indicate that it is. Discoveries in psychology and neurosciences demonstrate that pre-existing beliefs influence our thinking more than new facts because pre-existing beliefs have already been accepted and are found to be emotionally comforting. Denial avoids the emotional discomfort that accompanies cognitive dissonance and creates a clear path to continue a guilt-free lifestyle that contributes to global warming.

To avoid cognitive dissonance and the resultant discomfort of facts that we do not like, people engage in "confirmation bias" in which more weight is given to facts or beliefs that reinforce our cherished beliefs than those that do not. A related psychological dynamic is "disconfirmation bias" in which more weight is given to beliefs that seemingly disprove or debunk facts that we disagree with.

Confirmation bias is best illustrated by a Michigan republican. Her congressman, Justin Amash, had recently broken with the Republican Party and even called for the impeachment of Trump before the Democratic controlled house so voted. She had attended a town hall meeting to hear about his decision. She was interviewed following the meeting at which time she said, and I paraphrase here, "I didn't know he (meaning Trump) had done and said all that stuff." Her news outlet of choice was Fox News which had not sufficiently informed her of all of Trump's activities. The majority of people, both Republican and Democrats, select news sources that deliver the message they want to hear, the message they are emotionally comfortable with.

Confirmation bias is the tendency to search for, interpret, favor, or recall information in a way that strengthens or confirms one's personal beliefs or biases. Confirmation bias is seen in pushing away threatening information and pulling friendly information close. This effect is especially strong when one is seeking a desired outcome such as avoiding cognitive dissonance or protecting one's own self interests. Emotionally charged issues or deeply entrenched beliefs, such as those related to global warming, strengthen our desire to seek confirming opinions or news that supports our convictions. In other words, we gravitate toward those news outlets that favor and support what we already know or believe. While it is true we all are guilty of confirmation bias to some degree or another, when so much is at stake, as is the case with global warming, we need be watchful of that tendency.

Confirmation bias will cause us to think we are reasoning when we are actually rationalizing. When our "reasoning" is more about "winning our case" and not admitting we might be wrong, you can rest assured it is likely to be more rationalizing than reasoning. This type of reasoning is called "motivational reasoning" where our motive, such as winning our case, influences not only the sources we use in our search for "truth" but also what we will accept as "truth."

Another psychological dynamic closely related to "motivational reasoning" is "emotional reasoning" where we accept or reject facts based on how the facts make us feel. Facts creating emotional discomfort tend to be rejected. Facts that make us feel good tend to be accepted. Many of our opinions and decisions are emotion-based not fact-based. Many live by the motto of the Sixties: "If it feels good, do it." In this case, we could say, "If it feels good, believe it."

Our response to global warming is influenced by another psychological phenomenon two researchers, Troy Campbell and Aaron Kay (2014), call "solution aversion." They define solution aversion as the tendency people have "to resist or deny the problem when they are averse to the solution." That was witnessed during the coronavirus pandemic when thousands of young people were partying on the beaches in Florida and Mexico or when people refused to social distance or wear a mask when they went out into public. Solution aversion is also evident among people who resist or deny anthropogenic global warming.

I suspect most of us would agree that the solutions to global warming are not desirable. Solutions to global warming affect so many aspects of our lives adversely. To reduce the amount of CO_2 we release into our atmosphere requires we vastly reduce our use of fossil fuels. That would require we change our lifestyles drastically—how we heat and cool our homes, what and how we drive, the food we eat, our use of leisure time, our recreating, and so many other things we have come to expect and enjoy. Our lives simply would not be the same if we all took global warming more seriously.

We would have to sacrifice our self interests for the sake of others--others not even born yet. Sacrificing is difficult especially when there does not seem to be any compelling self interest to do so. One thing the coronavirus pandemic taught us, however, is that people are willing to sacrifice and even risk their own lives when they consider it necessary or their duty. Many heroes emerged during the coronavirus pandemic. And, yes, of course, the coronavirus pandemic brought out very selfish behaviors too as when people hoarded more than they needed or tried to make a profit off of life-saving items such as gloves, gowns and face masks. Yet, it is also true that many people are basically unselfish when the chips are down. When we know we must change and adjust, many have the strength of character to do so.

The trouble is that many people do not see the chips being down when it comes to global warming. In general, we often tend not to face problems until they are staring us in the face. If a problem can be ignored, it likely will be as we witnessed many doing during the coronavirus pandemic. In spite of the warnings and rising daily death toll, many went on with life as usual. In addition, it is hard for many to see very far into Earth's future when they are struggling with every day issues like meeting the monthly bills or dealing with a serious health issue. This is understandable. No one wants to add another problem, another stressor, to their list if they don't have to.

This tendency to protect our self interests is a necessary part of our human nature and actually essential for our survival; thus, we should not judge ourselves harshly. As noted above, many can put aside their own self interests and be heroes when the situation demands their very best. Our political, religious, economic and personal self interests are important to us and are hard to abandon or give up.

It is when these powerful motivators prevent us from considering the common good and facing the truth about a matter that we should worry. When our self interests cause us to hold onto false beliefs and engage in destructive behaviors they have become a problem not only for ourselves, as in the case of global warming, but for others a well. Anthropogenic global warming challenges our self interests like few things do.

For example, politicians worry about being re-elected. Re-election requires towing the party line or at the least not taking positions that run contrary to one's electoral base. This is obvious everyday in Washington as this administration attempts to gut environmental regulations designed to protect us. Few Republicans object to the administration's actions against the environment. Fortunately, as the public begins to accept anthropogenic global warming, more and more politicians are shifting their position on global warming in response to its wider acceptance. Too many, however, are still adhering to the party line as set by the leader.

In addition, it is the interest of some, but clearly not all, evangelical, conservative Christian groups to deny climate change. Pastor Robert Jeffress, for example, a member of Trump's Evangelical Advisory Board, speaking on Fox News said: "Someone needs to read poor Greta [note the derision of Greta Thunberg here] Genesis, Chapter 9 and tell her the next time she worries about global warming, just look at a rainbow. That's God's promise the polar ice caps aren't going to melt and flood the world again." Acceptance of the science of global warming would elevate science to a position that threatens other conservative core beliefs like the assumed sinful nature of homosexuality or evolution of the species.

The reverend's comments represent a blatant example of global warming denial. His assertion that God is the final arbiter of Earth's and humanity's fate serves as an excuse to do nothing for far too many people. This attitude is a danger not only to our health and wellbeing but also that of the environment. This attitude that removes humans from responsibility for the planet must be challenged and exposed for what it is, a threat to ourselves and our planet.

A Pew Research Center poll (9/2014) showed that many evangelical Christians are aligned with conservative politics opposing the teaching of the theory of evolution in schools, gay rights and abortion. Since the founding of the Moral Majority by the late Rev. Jerry Falwell, there have been close ties between the two groups. Many align themselves today with Trump who has claimed global warming is a hoax. Among some who do believe in global warming are those who believe that global warming will have only moderate effects. To accept anthropogenic global warming would betray their conservative religious and political beliefs and risk their standing in that group.

It goes without saying that much is at stake economically if we move away from the burning of fossil fuels. That is why the fossil fuel industry did not sound the alarm back in 1982 when their scientists made them aware of the damage the burning of fossil fuels was causing to our atmosphere. Resistance to the belief in global warming is still strong today within that industry. The fossil fuel industry has much to lose if we curb our use of fossil fuel and convert to green energy such as wind or solar. Their scientists have much to gain monetarily by spewing their propaganda. Humans are clever and creative, to be sure, but when those very abilities are used in such a way that endangers the public, they have to be challenged.

Even though there are alternatives to the use of fossil fuels, the industry resists change because of the economic impact it would have on their investments. That resistance is understandable. I don't believe there are many who want the fossil fuel industry to go out of business entirely, at least not immediately, or for investors to lose all their profits. I certainly do not. But that industry must accept the reality that we cannot continue to rely heavily on fossil fuels any more. That industry must start preparing itself for a major decline in the use of fossil fuels if they are to remain solvent and relevant and if this Earth is to remain life-friendly. Their challenge is to become more responsible citizens of this world.

Unfortunately, personal egos and professional reputations are involved too in this debate about global warming. Many are making a career and a profit out of denying anthropogenic global warming. At risk is their academic standing as well as their livelihood. In addition, no one likes to be accused of being wrong about such an important matter as global warming. Most of us like to think that we are at least as smart as the other guy or gal, especially academics, and will defend to our death our favorite beliefs.

Our egos, our sense of our own smartness, are at stake which makes it so difficult to admit we don't know something that everyone else seems to know or to admit we are wrong.

Our egos take a hit when we don't know something others seem to know or when our beliefs are challenged or attacked. We don't like to have to change our minds. For some it makes them look uninformed, inadequate or less intelligent.

Knowledge is highly prized in our culture. Not to know something about or have a position on something as important as global warming is unacceptable. Since most people do not have or do not take the time to research the subject to have an informed opinion on the matter, many settle for a belief that makes them comfortable and serves their needs. Moreover, once our minds are made up, we resist changing them for fear we will look stupid or uninformed. Protection of that ego is greatly important.

In addition to the motives above, personality differences also help explain some of the reasons why people resist the scientific facts. Chris Mooney makes it clear in his book (2012) *The Republican Brain: The Science of Why They Deny Science-and Reality* that we differ in personality in the following ways:

1. We differ on how open we are to new experiences and new information.
2. We differ on how comfortable we are with ambiguity or uncertainty.
3. We differ on how much we are inclined to accept authority and group think.
4. We differ on how we process information, that is, the extent to which we think with our emotions and rely on intuition or whether our thinking is more a by-product of careful deliberation and objectivity.

The reader is also encouraged to read Thomas B. Edsall's essay (2019) "The Whole of Liberal Democracy Is in Grave Danger at This Moment." He does an excellent job of reporting on some of the conflicting cognitive research about how and why humans think the way they do. Applying this research directly to the resistance to global warming, makes the problem of global warming denial more understandable.

Regarding point number one, those who prefer the emotional comfort of sameness, who like things the way they are and dislike and even resist change, will not seek or welcome new information or changes in their lifestyle. Global warming threatens their intellectual and emotional comfort zones. They want to cling to the old whether it is old information, old beliefs, old habits, traditions or their usual ways of doing things. The science behind global warming is unsettling for them. To accept the science would create too much emotional discomfort for them.

Regarding point number two, humans need to know and to have certainty. We need to understand the how and why of things, which is mostly a good thing, of course. Consider the fact that the universal response to tragedy is the question, "Why?" For example, I heard a holocaust survivor once, when asked what she would like to say to Hitler, responded by saying she would like to ask Hitler *why* he did what he did. We have to try at least to make sense of our world. It's when our world no longer makes sense that we get confused and anxious.

In addition, not only do we need to make sense of our world in the present, we need to know what lies in our future. We differ on how much uncertainty we can tolerate or be comfortable with. When we do not know what our future holds, we tend to worry and feel a sense of threat. The reality of global warming creates an uncertain future. For some it is better to avoid that uncertainty and remain in a comfortable delusion of all is ok.

Regarding point number three, some of us are just more conforming than the rest of us. There is a comfort in reliance on authority. Sometimes trying to figure things out for oneself can be just too difficult. Many find it much easier to let someone else do their thinking for them. The sense of comfort that comes with believing what some authority says strengthens our sense of loyalty to authority. Group think provides a comfort of its own too in that some find it better to agree than be at odds with the group. Conformity creates comfort. To know you are accepted rather than despised is reassuring. Religious groups know this. They know conformity can be enforced by threat of expulsion from the group or excommunication. We have such a need-- at times, a neurotic need-- to be accepted.

Daniel Kahneman in his book (2011) identified a type of thinking, he called WYSIATI. WYSIATI (what you see is all there is) refers to thinking that is fast and largely emotional, requires little effort on the part of the thinker and provides comfort. Such thinking is soothing, settling and quick. Because it feels so good, it also feels so right. WYSIATI creates no disturbing cognitive dissonance. Truth takes a back seat to comfort. Truth is what you already know. Facts, therefore, are irrelevant and are to be avoided since new information might disturb the emotional status quo.

In contrast there is a type of thinking we can call SDTT (slow, deliberate, thoughtful and thorough). This style of thinking takes more time and more deliberation, but it is more likely to lead to the truth or a better understanding of problems or facts. Even though such a search for the truth or the facts may be difficult, frustrating or even exhausting, one is in a better position to make decisions. High-quality decisions cannot be made without some hard mental and at times emotional work. With SDTT clarity and high-quality outcomes are more important than emotional comfort. Addressing the problem of global warming requires SDTT type of thinking. Unfortunately, the debate is often being framed by WYSIATI thinking instead of being guided by truth seeking.

What follows is a classic example of WYSIATI thinking. Over Spring break in Florida a reveler on a beach captured the attitude of many who deny global warming when she said during a live TV interview, "If I get corona, I get corona. At the end of the day, I'm not going to let it stop me from partying." WYSIATI thinking does not deal in facts. It's the old attitudes espoused by some in the 1960s, "If it feels good, do it."

WYSIATI thinking leads to a closed mind. A closed mind is a mind where facts do not make a difference. In such a mind, emotions rule. Facts are irrelevant. Such minds are like the Dead Sea. As the name implies the Dead Sea is dead. Located between Israel and Jordan at 1,412 feet below sea level, it is the world's lowest point of elevation. Because of that, water does not flow out of it. It is so salty that life does not flourish there.

Minds dominated by WYSIATI thinking are minds that are essentially dead; that is, rational thoughts will not flourish there. Feasting only on preferred thoughts and "facts" and taking in information only from narrowly selected sources, makes the mind sluggish and stagnant. New thoughts cannot grow in a mind that only allows input from select sources and never discards old ones. New ways of thinking or believing will not grow there. To grow a mind must take in new thoughts. If it holds onto the old beliefs and opinions it has always had, then it will languish. Nothing new can grow in an idle mind. The problem of global warming cannot be solved by closed minds.

To be open to growing our minds, i.e., changing what we believe, we must be aware of the motives behind our beliefs. If we find ourselves avoiding or rejecting established facts, then we must challenge our approach to facts. Motives that prevent us from accepting facts must be set aside so new information can get in. We must also understand how we process information, when we are engaging in confirmation bias, and what we do when we are experiencing cognitive dissonance. We must know ourselves, in other words.

There is yet at least one more explanation why people are not heeding the warnings of climate scientists about global warming. Psychological research strongly suggests that rebellious or non-compliant behavior is largely due to unconscious mental processes. The research indicates that unconscious thoughts, those thoughts we are not aware of, shape our decisions and opinions more than we realize. Underlying beliefs, attitudes and motivations that have accumulated over a life time of familial and cultural influences form what marketing psychologists call, brand connections. It is this intricate network of associated neurons or nerve cells that play a large role in our decisions and beliefs. The fact that much of our beliefs are of unconscious origins explains why they are so resistant to facts that are not consistent with our brand connections. Brand connections solidify over time and are resistant to change. A political party is an example of a brand connection.

I recognize the challenge of being a "connoisseur of self" as I suggested we become at the beginning of this chapter. Openness to new ideas or facts requires we understand our inner workings, i.e., how our mind works, what our motives are and insight into our unconscious mind. I am still working on self understanding and am learning something new about myself almost every day. I like how the Scottish poet Robert Burns phrased it: "Oh what pow'r the Giftie gie us to see ourselves as others see us."

Ask yourself, "How is the Earth seeing me?" Can you see yourself through other people's eyes? How open are you to new understandings about yourself? How honest can you be about yourself? Will you allow unfavorable information about yourself to emerge into your conscious mind?

A closed mind is a dangerous thing. No good decisions can come from a barren mind. Our Earth will only prosper if our minds open to the facts of global warming. If you find yourself with more insight into the reasons you hold the beliefs you do and more willing to consider facts you might have dismissed earlier, then your mind is growing and the Earth has another friend.

Hopefully, this chapter will help all of us be less judgmental and critical of those whose opinions and beliefs are different than our own. If we can understand the motives behind our own beliefs and those of others and what psychological mechanisms influence our thinking and believing, we have a better chance of mutual understanding and cooperation. Then we will improve our chances of addressing the problem of global warming. Humility is necessary for learning, but one must not be so humble that one fails to challenge falsehoods.

We need to keep John Stewart Mill's essay "On Liberty" in mind as we think about openness to new information, especially global warming in our case. He wrote: "The whole strength and value, then, of human judgment, depending on the one property, that it can be set right when it is wrong, reliance can be placed on it only when the means of setting it right is kept constantly at hand. In the case of any person whose judgment is really deserving of confidence, how has it become so? Because he has kept his mind open to criticism of his opinions and conduct. Because it has been his practice to listen to all that could be said against him; to profit by as much of it as was just, and expound to himself, and open occasion to others, the fallacy of what was fallacious."

The challenge with global warming is to have an open mind. Sound environmental policies will only emerge when men and women abandon falsehoods and embrace truth. Scientific facts, the means for correcting falsehoods, are only useful the extent to which we accept them. All of us must be open to criticism and change our point of view when the facts prove necessary. The great hope for this planet is our willingness to accept science and allow it to inform our response to global warming.

I want to close with a quote from the Buddha: "Do not believe what you have heard. Do not believe in tradition because it is handed down many generations. Do not believe in anything that has been spoken of many times. Do not believe because the written statements come from some old sage. Do not believe in conjecture. Do not believe in authority or teachers or elders. But after careful observation and analysis, *when it agrees with reason and will benefit one and all, then accept it and live by it.*" (italics mine)

Discussion Questions

1. What are your thoughts about Festinger's quote at the head of the chapter?

2. How did this chapter help you understand why you believe what you believe?

3. Explain how emotions have played a role in decisions you have made or beliefs you hold?

4. Share an example when cognitive dissonance applied to you? Are you experiencing any cognitive dissonance with the facts of climate change?

5. Have you ever witnessed a time when someone took a fact and twisted it to suit his/her own purpose? Have you ever done the same, for example, just to win an argument or settle a disagreement?

6. Have you ever made a decision to avoid an unpleasant result, what psychologists call solution avoidance?

7. Consider Chris Mooney's four personality traits. Do any of those apply to you?

8. Are your decisions or beliefs influenced by any "Brand Connections" such as political or religious?

9. How many of your beliefs meet the test put forward by the Buddha?

5

Where Are Abraham's Children?

> *"The environment is not just another issue, but an inescapable challenge to what it means to be religious."*
>
> - The Very Reverend James P. Morton, Dean,
> Cathedral of St. John the Divine

Our moral and spiritual understanding of the issue of climate change will begin with an important step back into our early history as a nation. Our founding fathers feared we would arrive at a point in our emerging democracy where the inability to agree and reach consensus would actually threaten our fledgling democracy. Little did they know how that inability would someday even threaten the very planet we live on and the quality of life of our children and our children's children.

In his farewell address, George Washington warned that the "spirit of party" would be democracy's "worst enemy." He went on to say that party loyalty "agitates the community with ill-founded jealousies and false alarms, kindles the animosity of one part against another...." That is clearly evident today in the debate about climate change.

John Adams also expressed concern about our young democracy. He wrote that the "greatest political evil" under a democratic constitution was the emergence of "two great parties, each under its leader, and concerting measures in opposition to each other." That political divide is revealed in poll after poll where Democrats accept climate change data more than Republicans.

Unfortunately, the issue of climate change has fallen victim to our two-party political system. Washington and Adams must be turning over in their graves with today's fight about climate change. A battleground pitting political parties against each other, just like our founding fathers feared, has ensued since our founding as a nation. Because of partisan bickering, Congress has been unable to enact adequate safeguards to protect our planet from pollutants and their harmful consequences to our environment.

Complicating matters even further, and equally detrimental, is the fight against all things science by many who are threatened by scientific facts-- a phenomenon our founding fathers could not foresee. The science behind climate change is denied by those furthering other agendas. False prophets have arisen within the scientific community claiming climate science is faulty. Many use that dissention to confuse the public and prevent any effective action on climate change even though over ninety percent of climate scientists agree on the facts of climate change and the human contribution to it.

Many challenges have faced our democracy since its beginning some two centuries ago. Some we have settled while others remain. The Civil War (yes, it took a war!) settled the issue of slavery even though blacks were still not treated as equal citizens until much later--and even now racial prejudice is alive and well in America. At least five and I'm sure there are more, contentious moral and spiritual issues still remain begging for workable solutions: abortion, immigration, LGBTQ rights, the teaching of evolution, the death penalty and climate change. Each of these issues creates much passion and has devoted followers. Effective solutions are hindered by partisan bickering and religious dogma.

Such issues showcase the fears of our founding fathers that we would be so aligned along party lines that we could not effectively solve problems. Somewhere along the arc of our brief history as a democratic nation, concern for the common good and the willingness to compromise was lost! We forget that America was envisioned by our founding fathers not just as a country or nation but as a cause promoting "life, liberty, and the pursuit of happiness."

If we are not able to think beyond politics and see climate change for the moral and spiritual issue it really is, we will never be able to unite around efforts to stem its tide. As stated before, it is disturbing to know that on the eve of the 2018 mid-term elections 78% of Senate Republicans still were denying climate change. There is too much at stake to remain divided on this critical issue. Political division is hampering our efforts to enact legislation to combat it. This delay only compounds the problems associated with climate change.

We have seen how making a crisis a political issue affects our response to a crisis. Our coronavirus pandemic was initially called a Democratic hoax by the President. He ignored the warnings of scientists about the danger of the spread and the importance of testing. The result was delay in our response to this deadly virus. He has also called global warming a hoax.

Once again, our response to global warming is less than it needs to be because it is not being taken seriously by lawmakers. Playing politics with a crisis is just plain nearsighted and deadly! And make no mistake about it, global warming will be a crisis. I for one would argue it is already a crisis for many throughout the world. The coronavirus pandemic should teach us that the best time to respond to a crisis is in the early stage of the crisis. Unfortunately, we are not applying that lesson to climate change!

Climate change is certainly a crisis and it is also one of the most, *if not the most*, compelling moral and spiritual issue of our time! The duty of all those who claim to be religious and spiritual to be, as the Lord Jehovah commanded, is to be good stewards of the Earth. We are to protect it and keep it healthy for others while we love all its inhabitants. This is our sacred duty.

Walter Brueggeman, noted Old Testament scholar, wrote (2018): "The prophetic tasks of the church are to *tell* the truth in a society that lives in illusion, *grieve* in a society that practices denial, and *express hope* in a society that lives in despair." Truth telling about the environment is required of religious people and religious institutions in my opinion.

In this context, I ask the question, "Where are Abraham's children?" Where are the Christians, Jews and Muslims in this climate change debate? Emperor Nero is falsely accused of fiddling as Rome burned for six days (CE 64). The point of the apparently false story is that as emperor of the Great Roman Empire he did nothing to save the city. Is it fair to say that Abraham's children are "fiddling" as Earth "burns?"

Much has been written by all faiths calling for an informed position on global warming and action to address the problem. A quick Google search will reveal the extent of the discourse and the call by all faiths to be better stewards of the Earth. Many within the religious institutions have urged people of faith to take climate changes seriously. From my limited exposure to the literature, however, I hear the voices of the clergy and scholars more than I hear or see the collective voice and action of lay Christians, Jews and Muslims calling for more responsible stewardship. I believe it a fair observation to say that Abraham's children, as a whole, need to stop being so timid on this issue. Many are certainly not timid speaking out about abortion or homosexuality, for example.

Please understand my intent is not to judge or criticize, though I recognize it may sound that way. Believe me I've got blood on my hands too. Read these words not as a blanket indictment of religion or religious institutions because I recognize that their collective presence is a good thing. I also get it when critics point out, and rightly so, that Abraham's children are responsible for much human suffering too. Abraham's children are for the most part good, decent people. I am happy to be associated with them as one who believes in something or someone bigger than myself.

Allow me to digress for a moment. After a sailing trip where Einstein had plenty of time to think deeply (could Einstein think any other way?) about God, he delivered his "credo." He said this in regards to being religious: "The most beautiful emotion we can experience is the mysterious. It is the fundamental emotion that stands at the cradle of all true art and science. He to whom this emotion is a stranger, who can no longer wonder and stand rapt in awe, is as good as dead, a snuffed-out candle. To sense that behind anything that can be experienced there is something that our minds cannot grasp, whose beauty and sublimity reaches us only indirectly: this is religiousness. In this sense, and in this sense only, I am a devoutly religious man." I also like Paul Tillich's idea of God as "the ground of being." I too stand in awe of this "Ground of Being."

Abraham's children are many and varied in their beliefs, hence so many houses of worship and creeds. Beliefs are not lacking. What I see lacking among Abraham's people as a whole is a passion for saving the Earth. Abraham's children bear a special responsibility for the Earth's wellbeing. We are the ones charged to take care of it, to be good stewards. We are the ones who need to, as Saint Francis of Assisi put it: "Preach the gospel at all times, when necessary use words." In other words, our actions, or lack thereof, are our message to the world.

As a member of the collective body of the children of Abraham, my aim is to offer an objective analysis. My desire is not to condemn but to challenge and inspire. I hope these tough words will promote discussion within the religious community and inspire action. Taking a stand on climate change is their opportunity to get on the right side of history and morality and their chance to "preach the gospel" of stewardship.

The children of Abraham represent approximately 4.2 billion of Earth's 7.3 billion inhabitants. Think of the force they could bring to bear on the problem of global warming and climate change if they all stood up in defense of God's Creation, our special but fragile planet. After all, their God commanded them to be stewards of the Earth (Gen.1:15) not to pollute it and destroy it to the point that it becomes almost uninhabitable. After all, the Apostle Paul wrote to Timothy: "For the spirit God has given us does not make us timid; instead, His spirit fills us with power, love and self control" (2 Timothy 1:7).

The time has long passed to be debating whether global warming, a fact the vast majority agree on, is due to human activity or not. The science is in and there is no doubt in any respectable scientist's mind that human activity, not some other factor, is the cause of global warming. To hold onto a contrary belief either reveals a lack of knowledge or bias much as we saw during the tobacco wars of the 1950s when the tobacco industry was able to confuse the public about the real effects of tobacco on our health.

Many of Abraham's children, as well as other Americans, of course, conveniently dodge the morality of global warming by falsely claiming that global warming is a political issue. Let me be clear, global warming is fundamentally a moral and spiritual issue not a political one. Those who claim otherwise are failing to distinguish the difference between policy and politics. Global warming is not a liberal vs. conservative issue or a Democrat vs. Republican issue unless, of course, those who want to maintain the status quo choose to make it so. To treat it as a political issue is a dodge which only hastens the day "Armageddon" arrives for future generations.

Global warming is a moral issue because it is an evil force in God's Creation, ruining God's work that He proclaimed as "good" (Gen.1:10). As the Earth continues to overheat the lives of countless peoples will be adversely affected. In fact, we are already bearing witness to the devastating effects of rising and acidifying seas, super-charged storms, prolonged droughts, unrelenting heat waves, and fouled air. Even if we now stopped burning fossil fuels altogether, enough damage has been done to our thin atmosphere that the effects of climate change will be felt for generations to come. I maintain that to continue on our merry way as if we do not have a serious problem already is immoral.

The Lord God expected His people to use the Earth's resources to be sure, but God did not expect them to ruin it for others who will follow. No, we were to protect our planet and nurture it so that its resources, its blessings, would make our lives and the lives of future generations better not worse. We are literally destroying God's Creation. We are changing it in ways that puts life itself in peril. The wonderful Earth generations have enjoyed is not the same Earth we are leaving our children. We need to be mindful of this Native American Proverb:

"Treat the Earth well. It was not given to you by your parents; it was loaned to you by your children. We do not inherit the Earth from our Ancestors; we borrow it from our Children."

Global warming is a moral issue on yet another count. We are commanded to love our neighbor as ourselves. Jesus went on to clarify to his disciples that we are to "feed his sheep," (i.e., take care of each other). Instead, we are in the process of making life miserable for our brothers and sisters. To condemn generations that will follow us to an Earth that threatens their lives like we have never seen before is immoral. But that is exactly what we are doing as we continue to contribute to global warming. We are condemning future generations to misery as we continue our unsustainable habits. That is immoral.

Abraham's children must become as concerned and passionate about this moral issue as they are issues like abortion and LGBTQ issues. Global warming is a moral issue on par with any other that occupies our minds and hearts. Just as people strongly protest abortion and act accordingly, they must begin to protest and act accordingly about global warming as well.

As the Native American proverb makes clear we did not inherit this Earth from our ancestors but borrowed it from our children and our grandchildren. Borrowing from our children is not something new to adults. We are borrowing from them now in respect to our unsustainable national debt and the rapidly depleting social security trust fund. We are living in the present with little to no regard for the consequences of our actions on future generations. In that respect, we are failing God who expects so much more from us. God must see us as selfish and narcissistic even if we do not see ourselves that way.

Scientists tell us the time to act is *now*! Further increases in carbon dioxide in our atmosphere will only result in more danger and harm to our children. We have a moral obligation to get global warming under control sooner rather than later. The moral obligation of all of Abraham's children is to get informed and involved.

Think about this: Who would be foolish enough to continue to run a car that was overheating? The engine would be ruined in a few minutes if we did not stop and address the underlying problem. If the car was not the driver's, as this Earth is not ours alone, it would be immoral to continue to drive the car since such abuse would clearly ruin it. Such is the dilemma we face now as we continue to "run" our lives in ways that increases the surface temperature of the Earth thus ruining it now and for centuries to come.

Fundamentally, the threat to our children's children is not only increasing levels of carbon dioxide but also indifference, ignorance, bias, silence and inaction. Imagine how things would change rapidly if 4.2 billion people joined voices and said, so all the world and all politicians could hear, "Enough is enough!" We have a choice and we must decide now. Will we be "good and faithful servants (stewards)" or not? The health and wellbeing of future generations depends on our response.

I hope for the day when global warming is considered as important as other moral issues like abortion, LGBTQ rights, and prayer in public schools, for example. These issues energize people and force them to take a stand. We must see the same energy and action when it comes to global warming. As in the old Up With People song, "Which way America, which way America to go?" The fate of humanity is in our hands now as it was during the cold war. Reason prevailed then; let's make sure reason prevails now.

Climate change is a spiritual issue as well. Perhaps I am splitting hairs here, but I believe morality and spirituality are separate but certainly related issues. Morality has to do with right and wrong or good and bad. Morality involves judgment. Global warming is judged to be bad or wrong because of the harm it does. Spirituality, as least as I define it, is not about judgment. Spirituality is about a way *perceiving* the world and *being* in relationship with humanity and the world itself--characterized by love, compassion and gentleness. Think of the Apostle's Paul's "Fruits of the Spirit" (Galatians: 5: 22-23). Spirituality, as defined here, transcends right and wrong. It transcends judgment. Take the spirituality quiz in Appendix "A" for a better understanding of spirituality.

The concern of spirituality is not the rightness or wrongness of our relationship with the Earth and each other, for example, but the essence or fundamental nature of those relationships. Behaviors in a relationship can be judged as good or bad, making those behaviors a question of morality. For example, it would be wrong of me to dump my used oil into the stream near my house instead of taking it to be recycled. I would never think of doing that, not because it would be wrong, but because that is not the kind of relationship I choose to have with my Mother, the Earth.

I want us to think beyond good or bad. I want us to be in a relationship with the Earth and each other that is more than just right or wrong. Morality is dependent on so many factors. Morality is relative to the person and situation while spirituality is not. Spirituality defies measurement or judgment. If we see other people, all of humanity, as the sacred beings they are, each worthy of dignity and respect, and each as valuable as any other, that is spiritual. If our being is one of loving compassion for all of humanity and other life forms and our presence on this planet is a gentle one, e.g., leaving as little carbon footprint as possible, that is the essence of spirituality. Such a relationship with the Earth, I maintain, is spiritual not moral.

Spirituality demands that one walk kindly and gently upon this Earth and among its inhabitants. Spirituality requires we consider the common good and in so doing be willing to make sacrifices for the whole of humanity. Spirituality means we change our life styles such that the Earth and all its inhabitants benefit not just ourselves. If we truly love our planet and all its inhabitants, then we do what is necessary to save this planet for future generations.

Our children and our children's children deserve better. Believe me Abraham's God is not happy about the way we are treating His creation. Read the story of Noah (Genesis 6 and 7: 1) where it is recorded God got so fed up with immorality that He decided to destroy all humanity. Had it not been for one good man, Noah, God would have done so. How is God feeling as He witnesses us destroying His wonderful creation? Is God finding one good man/woman? I believe God is finding many good people defending God's Creation. Read also the story of the Good Samaritan (Luke 10: 30-37). Imagine Earth is the one needing help instead of a person, a man robbed and beaten. Will you pass by on the other side so you won't have to deal with "him?"

The challenge to Abraham's children is captured succinctly by Alan Watts (1964), British philosopher and proponent of Eastern Philosophy: "Faced with this crisis, the egocentric style of personality will merely turn brutal, and there will be a frightful struggle for the survival of some elite group." We are witnessing such a struggle as the fossil fuel industry and its factions show disregard for how they are affecting the environment and those, because of their economic condition, most adversely affected by our changing climate. Written in 1964 his words were prophetic.

Watts saw Christians (and by extension, all of Abraham's children) as the solution to this problem. With Christians' emphasis on love and the welfare of others, the "unmindfulness" of technology with its ecological impact could be successfully challenged. He wrote of the "theocentricity" of Christians and its emphasis on the unity of man with God and each other as the perfect remedy for the greed of those who would destroy our planet. Jesus said it plainly, "…that they may be one just as we are one--I in them and you in me-- so that all may be completely in unity" (John: 17:23).

Global warming is a perfect opportunity for Abraham's children to silence those critics who claim that religion is irrelevant in today's society. The Earth is crying for help. Will you stop and minister to the Earth? God requires good stewardship of His creation. We must not be found guilty of being complicit in destroying God's creation by being complacent.

Discussion Questions

1. Do you agree with the Very Reverend James P. Morton's statement at the head of this chapter?

2. Has history, recent and past, proven Washington and Adams correct about a two-party system?

3. What is your reaction to the author's statement that playing politics in a crisis is just plain nearsighted and deadly? Is the issue of climate change caught up in politics? How so?

4. The author writes that climate change is perhaps the most compelling moral and spiritual issue of our time. Do you agree or disagree? What do you think might be more important? Why?

5. Are Abraham's children doing enough to combat climate change? What more could they do? Do they have any more responsibility for the Earth than anyone else or any other group?

6. What actions or attitude would characterize a spiritual relationship with the Earth?

7. Are you or Abraham's children as a group being complicit in destroying God's creation by being complacent?

6

There Is No Place Like Home

"Earth is the only world known so far to harbor life. There is nowhere else, at least not in the near future, to which our species could migrate. Visit, yes. Settle, not yet. Like it or not, for the moment the Earth is where we make our stand."

-Carl Sagan, Astronomer

He was sixty-four years old and in excellent health, until.... One day, after working hard on a landscaping project, he was so tired he could barely move. Two days later he was diagnosed with MS, multiple sclerosis. Since sixty-four is well beyond the usual age for a MS diagnosis, imagine his shock and concern. Until then he had been an active biker, skier, swimmer and hiker. He had also enjoyed playing full-court basketball with the young boys/men at the Y. Until that fateful day he, like most people I suspect, had taken his excellent health for granted. He never imagined he would have to face such a health crisis.

 I believe it is human nature to take our blessings for granted. We come to expect they will continue without any action on our part. Consider the case of so many couples I have seen in marital counseling over the fifty years I have been practicing. A quite common presentation is one in which he finally agrees to go to counseling after she says she wants a divorce. This is often after she has found another man who will listen to her. Often she has told him before how unhappy she was in their marriage. He had just taken their marriage and her love of him for granted, thinking she would stay with him "until death do us part."

The regular and familiar are often taken for granted; it is not until they are no longer present emotionally and/or physically or somehow threatened that we realize how much we really appreciated them. Consider, for example, perhaps the world's most famous cathedral, the Notre Dame Cathedral, that stood majestically for centuries in Paris. The world reacted in shock and horror as it burned in April of 2019. Its destruction woke many of us up to how much we valued and even revered it. Imagine how the world would react if, God forbid, the interior of the Sistine Chapel somehow was destroyed. Imagine the Chapel without its beautiful frescoes and tapestries that adorn its walls and ceilings. They were all painted or designed by the best of the Renaissance masters, most notably Michelangelo. Rome would still be a great city, but it would be less without it.

Sometimes it takes a tragedy to make us realize how important something is to us. A diagnosis of a life-threatening medical condition, for example, often makes people more appreciative of their health and causes a re-ordering of priorities. If we are not careful, we will create a crisis on Earth where life is extremely difficult. Some say we already have or are on the verge of doing so. Let us hope that it does not take a crisis to wake us up to the damage we are doing to our planet.

If we are honest with ourselves, we would have to admit that we take the Earth for granted. How often do we stop to appreciate the fact that the sun rises every morning as it has for the last 4.5 billion years; the fact that most of us enjoy an abundance of fresh, pure water just by turning a handle; or that our soils give us an abundance of food and that we use Earth's metals and petroleum products as raw material for the goods we buy? The common place is often not noticed and appreciated. We forget that many do not share in our blessings and that early humans had to literally fight for survival and struggle to live.

Humans have been living for the last couple of centuries like the Earth would always provide us with a livable atmosphere. Of course, it wasn't until the last few decades-- or, in all fairness, the last few years for most of us-- that we realized the damage our lifestyle was doing to the Earth. We have only recently begun to realize that we are ruining our planet, that we are destroying it for generations to come. Sometimes it takes a near tragedy (some would call it a tragedy already) to wake us up to the fact that we simply cannot afford to take our Earth for granted anymore. We now know that we have to protect our special planet, our only celestial home. There is no place in our solar system that is as friendly to life as planet Earth. If we ruin this place, we have nowhere else to go!

With global warming, we are learning that our actions on Earth have negative consequences for our water supply, atmosphere, and soil--all the things necessary for our life and comforts. Global warming is making believers out of most of us that our Earth is in danger. It is our wake-up call. Fear for its immediate as well as long-term effects is forcing us to reconsider business as usual. Scientists are warning us that we must stop using or drastically reduce our use of fossil fuels if we hope to avoid the worst consequences of climate change. Our neglect of our Earth has put us in an untenable position. We must choose between the health of our planet or the continuation of an unsustainable life style. We no longer can have both.

If we are to leave our planet safe for generations to come, we must change the way we think about the Earth. Our whole attitude toward Earth must change and when it does, we will begin to treat Earth with, may I say, the respect it deserves and even requires. We must come to love and appreciate the magnificent planet we have. When we love and appreciate someone or something we value them. We guard and protect the people and things we appreciate.

For centuries, we have not really appreciated our Earth. We have taken what we wanted from the Earth much too often without regard for the consequences of doing so. We have raped the land and polluted the soil, air and waterways. We were like someone with unlimited money on a spending spree who has no limit to what he can spend. We took what we wanted and assumed we could always keep doing what we had been doing without any harm to our environment.

Shel Silverstein's book *The Giving Tree* is a parable for our times. In the story the boy/man took and took from the tree until she had nothing left to give except a stump. He had taken without regard for the tree and she, out of unconditional love for him, had given all that he asked. Would it be too odd to think of the Earth as loving us as the tree loved the boy? I think not. We are exactly like that boy/man. We have asked much of Earth and she has indulged us. We have hurt her through our actions, however. For our own sake we must stop hurting her. We are "stumping" the Earth as the boy/man did to the tree. Now is the time to come to her defense and protect her from our worst habits.

Perhaps if we understood more fully that this planet is an exceptional planet among all those we know, then we would be more appreciative of its many blessings. Perhaps if we understood better how Earth and life itself came to be, we would value them and protect them more. Perhaps if we understood how fragile our planet is, we would treat her with more kindness. Consider, for example, there is no known place in this whole universe, as of now at least, and certainly no place in our solar system, where a planet is as supportive of life as our planet Earth. Of the thousands of exo-planets that have been discovered beyond our solar system, none are yet thought to be as life friendly as our Earth.

We have so many reasons to appreciate our Earth. There are several factors that make our Earth so unique and life-friendly. Consider the fact that Earth formed at just the perfect distance from our sun. If it had formed closer, it would be too hot for advanced life forms like humans. If it had formed further away, it would be too cold for advanced life. Earth formed in what scientists call the "sweet spot" or the "Goldilocks zone" of our solar system.

Next consider the fact that we have the perfect atmosphere for human life. The right amount of oxygen (O_2) and carbon dioxide (CO_2), for example, makes life as we know it possible. Too much CO_2 and Earth would be too hot, a problem we are experiencing now. Too little CO_2 and the Earth would not have the blanket of CO_2 that is essential to trap the heat we need to survive. Other gases such as hydrogen and methane that use to dominate our atmosphere have been greatly diminished by natural forces acting over millenniums of time. We also have an ozone layer that offers us protection from harmful ultraviolet radiation.

This perfect atmosphere is due, in large part, to Earth being a planet friendly to plant life. Plant life, like human life, requires liquid water. Too hot and water would evaporate and become gaseous like steam. Too cold and water would become a solid like ice. In either form, gaseous or solid, advanced life would not be possible. Many plants also require soil in which it can obtain the proper elements and minerals for growth. Our Earth after billions of years of evolution finally cooled down enough where solid soil could form. And due to natural forces, solid rocks broke down to form the soil with all the right ingredients that plants need to thrive.

Simple, one-celled plant life emerged some 3.8 billion years ago and were critical to the emergence of more complex life forms such as humans. Through a process called photosynthesis, plants take in CO_2 and make food, a simple sugar, and release O_2. It requires a chemical compound called chlorophyll to convert sun light into carbohydrates. Chlorophyll is one of those miracles of nature that makes life possible. Plants are life's primary food source. No plants, no food; it is that simple. Plants have also stabilized our environment by giving us just the perfect amount of CO_2 to retain heat in our atmosphere and O_2 for us to breath. All advanced animal life depends on plants to exist, either for their release of O_2 or as a food source. Without plants and their photosynthesis and without the miracle compound chlorophyll, we humans could not exist.

Much more could be written about the uniqueness of our planet, but this very brief account of Earth's specialness should suffice to deepen our appreciation of our celestial home.

Volumes have been written explaining how our Earth emerged to become a life-friendly planet from the original "Big Bang" some 13.8 billion years ago and how life evolved from a one-celled organism some 3.8 billion years ago to many multi-celled organisms found today on

Earth. I consider all this a miracle. I don't mean in the sense of something beyond science, but a miracle in the sense of the improbability of all this resulting in a planet that supports life and the presence of human life on this planet. The odds of such happening were not in our favor.

Two-time recipient of the Nobel Prize, scientist Albert Szent-Gyoergyi, in a lecture delivered for the "Symposium on the Relationship between the Biological and Physical Sciences" at Columbia University, had no better explanation for the miracle of life than to conclude, "...there is a 'drive' in living matter to perfect itself." His science could not explain this miracle in normal scientific terms so he had to resort to an unscientific word, "drive." Earth through billions of years of natural forces has perfected itself. Humans, through the process of millions of years of evolution, have perfected themselves. Hundreds, if not thousands, of textbooks have been written on these subjects. I only hope the brief account provided can help the reader more fully appreciate what I will unapologetically call the miracle of Earth and life.

Now let's turn our attention to another matter regarding our Earth, another factor that makes our planet unique--its raw beauty. Visit any other planet in our solar system and you will not find beauty that would match the beauty right here on Earth. Sure, other planets have their own beauty, like Saturn and Jupiter, for example, but none equal the beauty of the Earth. Anywhere one would travel on this planet, beauty can be found. Do a Google search for Earth's most beautiful places and you come away with a much keener appreciation of how beautiful the Earth is.

We don't always think of Earth as a beautiful place because, for the most part, we are not thinking of Earth much at all. This is understandable in light of the daily concerns we have-our kids, jobs, health, and so forth. The natural and man-made sights of beauty just escape our attention. Clyde Haberman (2019) writing in the *New York Times* about New Yorkers' lack of appreciation for the beauty of their own city, advised that people "Look Up" (the title of his essay). He wrote: "At any given moment, thousands of them are so focused on their screens that they fail to look up." He cites a host of beautiful things they are missing as they journey through New York as are denizens of any other major city whose eyes are glued to their screens. He writes, "What's the point of navigating the metropolis if you ignore the very sights that give urban life its verve?"

If we consider the sights that give Mother Nature her verve, we would have to plead guilty of not noticing enough. Who awakens each day, for example, with the Grand Canyon or some other natural beauty on his or her mind? Again this is understandable. I am merely pointing out that we need to give our Earth more thought as we spew more and more carbon into our atmosphere. We need to preserve the beauty of it and not fill it with our pollutants and use it as our dumping ground.

To appreciate Earth's beauty we don't have to travel to some faraway destination. Appreciating the Earth's beauty does not require that we go on vacation to take the time to see the beauty around us. Many of us live surrounded by the beauty of nature. We may live near a city park, in a well designed neighborhood, or near a forest or other sight of natural beauty. In the spring there are daffodils and crocuses to be seen heralding the birth of new life. The summer the whole Earth is covered with flowers. In the fall, beautifully colored leaves dominate our landscape and announce the end of another summer full of its own beauty. All it requires to see beauty is to have an appreciation for beauty and the eyes to see it.

Just like Einstein, I stand in awe of the universe but especially our Earth. Psychologist David Yaden, in an article written by Galadriel Watson (2020), speaks about two kinds of an awe experience. A *perceptual* awe experience occurs when one sees or experiences something that is highly unusual or out of the ordinary. A starry night will do that for me. A *conceptual* awe experience occurs when one imagines or experiences an idea, for example, that is so radical or mind-blowing. This could be an inspiring lecture or speech or a concept like "dark energy."

Let us preserve as much of this paradise as we can for those who will follow us. Even though it is too late to leave them the pristine Earth we have enjoyed, it is not too late if we act now to leave them a good measure of what we have enjoyed from our Earth. Earth is a sacred trust we have been given. Like any trust, it must be used wisely. We have failed to do this initially out of our ignorance but now out of our neglect. Earth is an investment we are managing for future generations. Let's invest wisely. My prayer is that we won't disappoint our children and their children since we now know the consequences of our actions.

While appreciating the beauty and uniqueness of our planet, we must not forget its inhabitants. Homo sapiens sit at the top of the evolutionary tree. There is not a single species more advanced than ourselves. We may not be able to fly like birds, swim through the ocean like dolphins, run as fast as cheetahs, or uproot a tree like elephants; but, we have invented flying machines that can go higher and farther, submarines that can go deeper and stay submerged for several months, cars that can travel over two hundred miles an hour, robots that can build cars, computers that can play chess and even think like humans, and heavy equipment and dynamite that can level mountains.

We are unique among all species. We are the culmination of life's some 3.8 billion years of evolution. Seven million years ago hominids, beginning with *Sahelantropus tchadensis*, started their march to becoming *Homo sapiens*. We need to celebrate our evolution from our very ancient ancestors and stand in awe of the evolutionary process that eventually gave rise to our species. Think of the odds of single-celled organisms evolving into trillion-celled organisms that can walk upright, engage in complex communication and travel beyond our own planet. We are indeed special among all life forms.

There is no animal more complex and unique as humans. We have built pyramids and castles, created unparalleled works of art, and developed means of communication unsurpassed in the animal kingdom. We have built long canals to connect oceans and have visited other celestial bodies millions of miles away. We have walked on our moon and looked back through thousands of miles at our own planet. Our accomplishments are unique among all species. If I may brag a little, we are pretty hot stuff!

But, with all our knowledge and accomplishments we have been careless with our special talents. We have not done and are not doing a very good job of taking care of our planet and each other. The evidence is clear in the rising temperature and changing climate of our planet and in our homicide and suicide rates, our endless wars we fight, addictions, starvation, premature and unnecessary deaths of thousands each year, and so on. Human life (like many, I would argue, *all life*) is sacred.

The words of Dr. Abraham Loeb, chair of the astronomy department at Harvard University summarize this discussion well: "The most fundamental lesson is simple. We must treasure all the good that nature gives us rather than take it for granted, because it can easily disappear." The lesson he was referring to is the lesson we need to take from the coronavirus pandemic. This is the same lesson we must apply to global warming.

I live close to nature, surrounded by trees, flowers, all manner of native wildlife, a small stream that makes music in the spring, and hills. Allow me to share a poem I wrote to express my appreciation of this planet.

> I arise at daybreak
> The sun is mine
> Like the bountiful Earth
> I too am kissed by its presence.
>
> I hear the babbling brook
> I pick berries from the vines
> To embellish my cereal
> Am I not as rich as a king?
>
> Birds serenade me each morning
> Chipmunks and squirrels amuse me
> Butterflies grace my flowers
> Reminding me of the wonders of life
>
> Kings have jewels and palaces
> Presidents have power and prestige
> Famous people are idolized
> Me, I am richly blessed.

One final note: The man diagnosed with MS is now 78 years young and enjoying riding an e-bike and still doing most of his other activities. Life is good for him again. One thing he does not do anymore, take it all for granted as if it will never end or that he won't have other challenges. His appreciation for life and health has increased immensely since his diagnosis. I should also add that he is also enjoying the challenge of writing this small book.

Discussion Questions

1. Do you take your blessings, your health, for granted? In other words, are you more aware of what you have or what you lack?

2. Do you take the health of the Earth for granted or do you, for example, feel blessed every day by the sun or the presence of green plants?

3. Do you believe we are borrowing the Earth from our children? In what way(s)?

4. What is some of the beauty you have witnessed directly or indirectly on this planet?

5. Are you missing the beauty and splendor of this planet in your daily life?

6. Is there a special place you find so beautiful, awe inspiring, or mysterious, that you would like to visit it again?

7. Do you consider all life and this Earth as sacred?

8. What do you make of the specialness of our planet in our solar system and even in the universe (at least as far as we know now)?

7

For the Love of Our Mother Earth

"This is what you shall do; love the Earth and sun and animals."

-Walt Whitman

"Even after all this time the sun never says to the Earth, 'you owe me.' Look what happens with a love like that; it lights up the whole sky."

-Hafiz, Sufi poet

If we loved the Earth like the sun loves the Earth, there would be no problem of global warming. I say love because love means to have someone's welfare at heart. Love means making sure they or it is well taken care of. Many say they love their home or their car meaning they feel an attachment to it and keep it in good shape. Love is just that, taking good care of something or looking out for someone's best interest.

It is strange how difficult it is for some to say, "I love you." Many of my clients have told me their parents never said they loved them before they died. That was so painful for them. Even when we know it we often fail to say "I love you." I find myself hesitating to say I love you to friends that I love but I find it easy to say "I love you" to family.

People often show their love more often than they say it. Deeds often substitute for the words of love. The sun and Earth, of course, can't say they love us, but they surely show it every day. This should not be a radical concept, Earth loving us. After all, when you say you love someone you want to show that love by your actions not just words. I say the Earth loves us because she provides for us and she never asks for anything in return. But we must take care not to ruin her precious gifts to us.

Earth's love for us knows no bounds. Look at how she has provided for us across the two hundred thousand years of Homo sapiens history. She has provided us with food, water, air, shelter, clothing, materials to build great mansions and homes, sights unimaginable to please our eyes, lakes, mountains, snow and ice for pleasure and entertainment, and countless other blessings without which our lives would not only be diminished but impossible. Is that not the essence of love? When you love people, you take care of them. You provide for them. You protect them.

Imagine if we loved the Earth so deeply. What a light we would shine on the Earth! Imagine if we had gratitude for her many blessings and actually showed it. How would that change how we treat her? Would we abuse her like we do? Would we make mountains out of our wastes or dump our wastes into her air and streams? Of course we wouldn't! Love does not hurt. Love does not take advantage. Love does not diminish.

Many of you may be thinking all this discourse about the Earth loving us is ridiculous. How can an inanimate object such as the Earth be capable of love? Ah, that's where you are mistaken. The Earth is not an inanimate object. She is a living, breathing organism! The whole of the Earth, her inanimate and animate parts, constitute a living being. I would argue that she is even a sentient being, that she has consciousness as many others claim. I'm not suggesting that the Earth is conscious of each and every one of us, but she is conscious of humanity, of life. In other words, she is awake. She responds to stimuli, our actions, for example. She is a living, moving being who is always changing, always growing. She has moods just like we humans. She is, of course, capable of actions just like we are. She breaths the very air we breathe.

Earth is a sacred being. David Brooks (2019) advises we see "the sacred in the realities of everyday." He writes that real living is like watching a play on two levels. On one level are the ordinary lives of the characters, what he would call the horizontal level. Then there is an underplay, the vertical level, where a sacred or spiritual story is unfolding. The same is true of our Earth. On the horizontal level there are sunrises, sunsets, rains, storms, etc. On the spiritual level, however, is the Earth's love for us unfolding before our very eyes. This love permeates the physical world, the world of our senses. He explains that Jews have a term, *tzimtzum,* to describe the way spiritual essences infuse the material world.

We need to cultivate a love for the Earth that is as strong as the Earth's love for us. Love would compel us to take better care of her. Love would compel us to protect her. How can such a love for Earth be developed? The same way we learn to love a child or an adult. We draw close to them. We engage with them. We spend time with them. We consider them important to us. We feel a sense of responsibility for them. We share with them. We want the very best for them. All these actions create a bond of love. They cultivate love for the person and they will cultivate a love for the Earth.

If you have ever cared for an infant, you know that it is the very helplessness of the child that calls forth our love for the child. Our love wants to insure the infant's comfort and survival. The helplessness of the infant, in combination with its responsiveness to our actions, creates in us a love for the child. In fact, we know that an unresponsive infant is at risk for neglect if not abuse. That child simply does not elicit the same loving, caring responses from the parent the child who is more responsive does. We need that responsiveness along with our sense of the infant's helplessness to complete the bonds of love we share.

Let's see how all this relates to the Earth. First of all, we must dismiss the notion that the Earth is not helpless and doesn't need us. Of course, at one level, both are true. The Earth was here long before we were and will be here long after we are gone--whenever that may be. Earth will survive whatever we do to it. However, for the health and wellbeing of the Earth and the humanity that occupies her, she needs our love just as we have needed hers for the 3.6 billion years life has been on this planet. Our actions affect her ability to care and provide for us.

We are in the process of fundamentally changing the health and wellbeing of the Earth as far as its capacity to support and nurture us is concerned. Before the wide spread use of fossil fuels, Earth was getting along just fine in her care for *Homo sapiens*. With the dramatically increased use of fossil fuels, starting with the Industrial Revolution, Earth began to suffer. Her capacity to care for life has been diminished over the last few centuries. She can no longer provide the wonderful conditions for us like she did before.

We have changed the Earth for the worst. The absence of love will do that. Look at all the people in prisons and all the youth in trouble and incarcerated because they were not adequately loved. Love is an absolute essential ingredient in the health and wellbeing of a child, an adult, and even the Earth. We know that, but we don't apply that knowledge to the Earth. We have been treating the Earth like she did not need anything from us. The result is an Earth that is in trouble, an Earth that can't provide for us like she once did.

Perhaps this all sounds like crazy talk to some of you. But, one thing is indisputable; the Earth is being hindered in her capacity to care for us like we need her to. Our actions are affecting her in ways that are making our lives more miserable and problematic. We are diminishing the capacity of Earth to give us the life and conditions we need and have heretofore enjoyed.

The solution, once again, is to love the Earth with a love we have not yet shown before. That is the only action that will preserve the Earth's ability to love us fully and save us from ourselves, our own careless actions. Love is quite often the only answer that makes sense, isn't it? Viktor Frankl, *Man's Search for Meaning*, wrote. "The salvation of man is through love and in love." To which I would add, the salvation of Earth and her inhabitants is through love and only love. Greater love of Earth is the only thing that will preserve Earth's capacity to provide for us.

Love for the Earth can be cultivated by engaging with her. Go outside on a clear, dark night, away from artificial lights, and look up at the canopy of stars overhead. If that doesn't deepen your sense of awe for the vastness and mystery of the universe, I don't know what will. Listen to the chorus of birds welcoming the end of darkness and the dawn of light. Walk barefoot through the dewy grass or on a sandy beach. Allow a gentle rain to wash away or minimize all your cares and worries. Put your bare feet into a pond or wade out to where the water meets your knees. Dive into a cool pond on a hot day. View the vista from the top of a high hill or mountain. Climb a tree if you are able. Hug one if you are so inclined. Make a snow angel or a snowman. Grow your own vegetables and fruit. Plant a flower. Pull a carrot out of the ground or pick a tomato off a vine. Go to an orchard and pick your own peaches or apples. Mow your own grass. In short, immerse yourself in nature. I promise you such actions will deepen your love for and appreciation of Earth and your desire and commitment to preserve her for future generations.

Alan Watts, the great British philosopher and promoter of Eastern Philosophy, believed in an *ecological* view of man. He taught that man and nature are inseparable and spoke of them as a whole, as organism/environment. He wrote (1964): "Ecology must take the view that where the organism is intelligent, the environment is also intelligent, because the two evolve in complexity together and make up a single unified field of behavior." In other words, even though we *feel* our existence separately, we and our natural environment are one. We are interdependent. What affects one affects the other. That, in essence, is what the environmental movement is about. We must come to realize that we are not separate from our environment. Moreover, just as we love others and ourselves we must love the Earth as well.

The late senator George McGovern and presidential candidate lost a daughter to alcoholism. It was not unexpected as she had struggled with alcoholism much of her life. His words about her life and death are instructive here. He wrote, *Terry: My Daughter's Life-and-Death Struggle with Alcoholism:* "Both her life and her death have taught me much. Perhaps most significantly she has taught me that life is not only precious, it is fragile and uncertain-and that we should love each other more. I wish that I would have held her closer...."

He had deep regrets about how he had withdrawn from his daughter in the last few months of her life. He did not abandon her, but neither did he care for her in ways that he wished he had. I hope we who are alive today do not ever have to regret the way we have cared for this precious and fragile Earth. There is still time to love our Earth more responsibly, more fully. WE should treat the Earth in such a way that our children will never curse us for the way we lived.

To quote Stewart Udall, Secretary of the Interior under President John F. Kennedy: "Cherish sunsets, wild creatures, and wild places. Have a love affair with the wonder and beauty of the earth." Fall in love with the Earth today as if you are falling in love with another person.

Discussion Questions

1. How would you explain love to a child?

2. Can the Earth be loved like a child?

3. Do you consider it silly to talk about loving the Earth?

4. Do you see the sacred in the realities of everyday life (David Brooks)?

5. What does David Brooks mean when he talks about the horizontal and vertical levels of life? Which level do you live on? How would living more on the vertical level affect our environment?

6. We know how the lack of love affects a child. Could it be said the Earth is being affected by a lack of love on the part of humanity?

7. If you had a love affair with the Earth as Udall recommends, what difference would that make in your lifestyle?

8

Can Global Warming Be Stopped?

"Nobody made a greater mistake than he who did nothing because he only could do a little"

-Edmund Burke

"Never doubt that a small group of thoughtful committed citizens can change the world; indeed, it's the only thing that ever has."

-Margaret Mead

As I am writing, the world's attention is focused on the coronavirus outbreak which started in Wuhan, China. Thousands of people have died from this disease so far and more are dying each and every day. The world worries as it spreads around the globe and within borders. It has become a pandemic. Governments are taking steps to prevent its spread. It is necessary for us to devote our resources to stopping this pandemic.

But, where's an equal concern for another scourge plaguing all of us which only gets worse year after year? Global warming is relegated to the back burner of our government's mind and even denied by those who should know better. If we would treat the threat of global warming more effectively than we did the threat of the coronavirus initially, we would be way ahead of the game instead of behind the eight ball. In global-warming terms, the Earth is the host organism, the virus is human activity, and the symptom is global warming. Whether global warming and the coronavirous are equally "deadly" remains to be seen, but both require vigorous action to stop them. I urge us to take global warming as seriously as we are taking the coronavirus.

David Wallace-Wells (2019), *The Uninhabitable Earth*, offered a disturbing opinion. "That so many feel already acclimated to the prospect of a near-future world with dramatically higher oceans, should be as dispiriting and disconcerting as if we'd already come to accept the inevitability of extended nuclear war—because that is the scale of devastation the rising oceans will bring." That is a sobering, but at least partially true, observation in my opinion.

The climate models presented by Dr. Irwin (see chapter 10) predict an Earth on which life as we know it will be changed. Depending on the model, some forecast a future that is less negatively impacted than others. All models agree the outcome all depends on human activity from this day forward. If we keep polluting and fail in our attempts to capture CO_2 from our atmosphere, our future is bleaker than if we do not.

Two climate writers who have been following this issue for years have reached different conclusions about Earth's and humanity's future. Emma Marris and Elizabeth Kolbert, writing in the April 2020 issue of *National Geographic*, present two different outcomes Ms. Marris sees the glass half full while Ms. Kolbert sees it as half empty. One is the optimist while the other is the pessimist. One is bullish and the other is bearish.

Drawing on her many years of studying the subject and from data compiled by Project Drawdown, Ms. Marris presents a future where we make the changes necessary to save Earth and humanity from a worst-case scenario. While acknowledging we cannot undo the damage we have already done to our Earth, she finds comfort in the growing interest of our young people, in changing public opinion and in efforts already being made to stop damaging our planet further. She writes: "But in the midst of a swirling sea of sorrow, anxiety, fury, and love for the beautiful weirdness of life on Earth, I find an iron determination to never, ever give up." For her, the future of our planet rests in this "iron determination" to protect our planet and save it for future generations.

Ms. Kolbert, on the other hand, does not forecast such a bright future for our planet. She is not as confident that we will do what is necessary to protect and improve our planet. While Ms. Kolbert sees the same changes in attitudes of the public and the increased interest among our young people as Ms. Marris does and while she acknowledges we have the knowledge and increasing technical expertise to stem the tide of global warming, Ms. Kolbert is less optimistic. The problem she sees is our ever increasing demand for more and more of all the things that has put us in the perilous position we are in. This incessant demand for more requires we use more and more fossil fuels to maintain our unsustainable lifestyles. The continued demands we put on Earth's resources will be our undoing in the long run she argues.

Ms. Kolbert quotes veteran NBC, TV broadcaster, Hugh Downs, who said at the opening of our first Earth Day: "Our Mother Earth is rotting with the residue of our good life. Our oceans are dying, our air is poisoned." Ms. Kolbert sees no evidence that this rotting will stop at least any time soon. I believe she would like to be more optimistic, but she sees a different reality than Ms. Marris. The evidence for our reducing our negative impact on our environment is lacking from her perspective. She concludes: "If current trends continue, the world in 2070 will be a very different and much more dangerous place...."

What will be Earth's and humanity's future? I do not know, but I want to be optimistic like Ms. Marris. However, I recently asked a high school environmental science class if they had heard of Greta Thunberg, thinking I would see a large showing of hands. Imagine my surprise when only three hands in a class of seventeen went up. I then asked them how many were concerned about the Social Security Trust Fund. As I expected, I got blank stares. Then I asked how many were concerned about global warming. Again only a few hands went up. Is this disinterest typical of our youth? I don't know, but it concerns me, as it probably does you.

Putting aside the question of Earth's and humanity's future, here is a question I want to ask the citizens of the world, "Had enough yet?"

GLOBAL WARMING: CAN IT BE STOPPED?

Hardly a day goes by that there is not some news item pointing to global warming. Almost every week we hear of unprecedented flooding, record breaking tornadoes and hurricanes, blizzard-like snow storms, intolerable and relentless heat waves, heavy rain storms, uncontrollable fires--most likely directly or indirectly related to global warming. When will we respond with the urgency this problem deserves? While we wait on our governments to do something, the problem only gets worse. We are like lemmings willfully following our leaders over the cliff. The time is now that we all get involved and take action.

Intergovernmental Panel on Climate Change (IPCC), a United Nations Agency, concludes: "Warming from anthropogenic emissions from the pre-industrial period to the present will persist for centuries to millennia and will continue to cause long-term changes in the climate system.... Climate-related risks for natural and human systems are higher for global warming of 1.5 degrees C than at present, but lower than at 2 degrees C (*high confidence*)."

In other words, through our carelessness, our "business as usual" lifestyles, we are putting our lives at risk and that of generations to come. It is imperative that we get our act together and act now to prevent the worst effects of climate change. Earth's and humanity's future depends on all of us.

Prevention begins and ends with realism. We cannot settle for half-baked solutions that have little or no basis in reality. Planting a trillion trees, for example, is simply not a realistic solution to the problem of global warming now and perhaps not even in the future. That was the recent proposal of the World Economic Forum--plant a trillion trees. Many governments immediately adopted the idea and then went to sleep. That idea sounds sensible until it is subjected to realism.

Jane Roberts (2020) reports that scientists with knowledge of climate change dismissed the idea, calling it "incorrect scientifically and dangerously misleading." Even if a trillion trees were planted, it would be years before they would be absorbing enough CO_2 to make any difference in our average global surface temperature. Meanwhile, if that were the only response to global warming, our planet would continue to warm. Many find planting a million trees an attractive solution, not because of its scientific merits, but because it requires little from governments to propose it. Now, let's wait and see how soon one trillion trees are planted. I hope they are serious because deforestation, the burning of forests in particular, is already a major cause of global warming.

Such a proposal, however, is not grounded in either science or realism. In this same category, *at least at this time,* are such proposals as carbon sequestration, fusion power, tidal power, heating salt, compressed air, "carbon farming," algae-based biofuels and geo-engineering. Some of these do hold promise, but as solutions that will have immediate and substantial impact they are unrealistic. Much more research and development must be done before any of these help reduce our carbon footprint.

We must not think we are doing anything substantial about global warming until we realistically address the problem. To reduce our GHG emissions, we must do two things and do them now-- use less fossil fuel by converting to alternate energy sources and consume less energy. These two ideas are not being taken seriously by our government or by most of the public.

A poll taken by Yale University and reported in the journal, *Environmental Research Letters* (2017) found that the average American would be willing to pay $177 per year or 14.4% more on their energy bill in support of a carbon tax. In addition, 80% of them would want that revenue to be used to develop clean energy and 77% would like to assist workers in the coal industry. Only 59% would want to reduce federal income taxes! This is great news. Many Americans are waking up to the threat of global warming. At the same time, alternative energy sources, particularly solar and wind, are becoming less and less expensive, but they still cost more. Just as with gun legislation, our politicians are once again are not in tune with the desires of the people.

For more ideas on how you can help reduce our carbon footprint, see the report in drawdown.org *which* ranks solutions to global warming. It is well worth a read for people who want to lower their carbon footprint. Another useful source for ways to reduce our carbon footprint can be found in real-leaders.com.

Experts make it clear that it will take massive government efforts to effectively reduce our carbon footprint. A great deal of R&D will be required to bring some of these ideas to fruition-and even then we will not know how practical they will be until we have tried them. Cooperation between government and corporations will be required like that was seen during WWII when all worked together to help defeat the enemy. We will all have to work together again with less concern about profit and more concern about finding effective solutions to benefit humanity.

We can accomplish great things when we work together with each other and other governments. For example, when the world became concerned about the potential loss of our ozone layer, governments banded together to stop the release of chloroflurocarbons. This collective action prevented further depletion of our ozone layer. Recent measurements find it to be rebounding from an earlier decline. We did it then and we can, with resolve, cooperate to protect Earth and its inhabitants again.

Kevin Drum (2019) offered this perspective: "The real issue is this: Only large-scale government action can significantly reduce carbon emissions. But that doesn't let any us off the hook. Our personal cutbacks *might not* (italics mine) matter much, but what does matter is whether we are willing to support large-scale actions… that will force us all to reduce our energy consumption."

To convert to green energy and to use less fossil fuel energy, these are the challenges we must be realistic about. Any less action will not leave this planet a good place for us or our children. According to the IPCC, CO_2 emissions are coming from these sectors: transportation (14%), electricity and heat (25%), industry (21%), building (6.4%), and agriculture, forestry and farming use (24%) and other energy (9.6%). These facts must inform our actions if we are to be realistic in our approach to global warming. Addressing the major CO_2 emitting vectors, transportation and power, would reduce our CO_2 emissions the most. That is where we must start if we are serious about global warming.

I am cautiously optimistic. I wish I could be more optimistic. I wish I could reassure you, the reader, that we've got it under control. The accumulating climate data just does not support a rosier picture. Human nature being what it is doesn't lend itself to optimism as we can really be self centered and not at all future oriented. On the other hand, consider how unselfish we were during WWII when so many were willing to sacrifice. Even today, people are making sacrifices everyday in service to humanity and their countries--police, soldiers, fire fighters, teachers, nurses, doctors and so many more. In addition, many others are already making changes in their life styles to help the planet.

Perhaps if we saw everybody sacrificing to save our planet for ourselves and future generations, we all would be willing to pay that little more that converting to alternative fuels would cost. I like to think so. The problem is leadership is lacking. On the other hand, many municipal governments, businesses and corporations are doing more even as our president pulled us out of the Paris Accord. What is needed, however, is leadership at the top of our government, the President and his cabinet. Instead, they are still minimizing and outright denying anthropogenic global warming. They are making the same mistake with global warming as they did with the novel coronavirous-- that is, deny, minimize and delay action.

There is still reason to be hopeful, however. More and more people are demanding their governments take this matter more seriously. In the US, strong leadership is beginning to emerge at the state and local levels. Public support for robust action is growing. WWII proved we can do it. The ozone problem also proved we can do it. Our response to the novel coronavirous proves what we can do when we take a problem seriously enough and cooperate with each other locally and internationally. Let's dedicate ourselves to the task of saving humanity from another threat to our existence from an evil force, which is what global warming really is.

But, here's another problem, at least here in the United States. A major roadblock to cooperation is the great divide in our politics. Until the novel coronavirous pandemic, at least, there was little agreement among Republicans and Democrats on matters of affecting all of us. Congress was able to agree on a robust course of action to address the economic hardship individuals and businesses faced due to the novel coronavirous. Whether we continue to cooperate or not, remains to be seen. In addition, there is wide disagreement on the cause of global warming. Certainly, no cooperative agreement addressing global warming is on the political horizon. In addition, the nations of the world are not working together for a solution. There is agreement on a goal limiting CO_2 emissions—take the Paris Agreement, for example--but much work remains to be done to achieve that goal.

Research has demonstrated that there are personality differences between the average Democrat and Republican (Chapter 4) that cause them to look at issues differently, but we must work through those differences if we are going to solve the problem of global warming. Here I am not optimistic. Until we elect leaders that make addressing global warming a priority, our response to global warming is not going to be robust enough to make a difference.

Paul Krugman (2019) offered this sobering analysis: "The politics of climate change have followed a similar trajectory. Global temperatures are setting records, while climate-related catastrophes like the Australian wildfires are proliferating. Yet a majority of Republicans in Congress are climate deniers--many of them buying into the notion that climate change is a hoax perpetrated by a vast international scientific conspiracy-and even those… who grudgingly admit global warming is real, oppose any significant action to limit emissions."

Who will step up and call foul? Who will challenge the political establishment? Where is the leadership on this issue? Sorry folks, but our politicians are making global warming a political issue. Let's not make the same mistake they are. All the more reason that people who love the Earth, who are convinced that climate change is real, who care about the fate of humanity now and in the future, who are willing to make sacrifices for people they don't even know, most of whom have not even been born yet, get out and vote and vote for people who will be willing to make the tough decisions necessary to save our planet. The most powerful action one person can take is vote and vote for climate-friendly politicians whether they be Democrat or Republican. What matters is not their political party, but their embracing the science of climate change and well- informed scientific solutions.

As we move forward in our campaign to slow the warming of our planet, we need some guiding principles. I offer the following for your consideration:

1. Treat the Earth well. We need it to be healthy. Our wellbeing depends on it.
2. The Earth was not given to us by our parents; it was loaned to us by our children. We did not inherit the Earth from our ancestors; we borrowed it from our children as a Native American proverb declares.
3. Love the Earth as you love your mother for like our mother she is a living, breathing organism that gives us life and takes care of us. She is the ultimate "Giving Tree." She gives and gives while asking only that we treat her well in return.
4. Human life is sacred; it should never be sacrificed on the altar of progress, profit, or personal or political gain.
5. All humans are not only sacred, they are equally valuable. No one person, race, gender or nationality is more important than any other. Though some may have accomplished more or have more riches than another, that only makes them more successful, not more important.
6. We all share the same planet. Though it is divided into nations and continents, the atmosphere and the oceans are not so divided. We all breathe the same air, share the same waters, and depend on the same natural resources. What affects one affects all. As the poet John Donne penned: "No man is an island…, every man is a piece of the continent, a part of the main…; And therefore never send to know for whom the bell tolls; it tolls for thee."
7. We are people with two bad habits which in the case of climate change put our lives at greater risk. We take "ordinary" things for granted and we put off doing things until the last minute.
8. Facts are facts. There are no "alternative facts." Facts are not bound by political or personal boundaries. There aren't conservative or liberal facts. Facts have no sides.

The practical application of these principles would require all of us to sacrifice for the greater good not only for this current generation, but for generations to come. We would be willing to pay more, for example, for heating our homes and driving our more fuel-efficient cars. Gross national product (GNP) would not be the measure of our success as a country but quality of human existence (QHE) would be. Concern for all citizens of this planet would replace narrow nationalistic impulses. We would consume less instead of being the throw- away culture we are accused of being. Like citizens during WWII, we would willingly conserve our precious resources, the most important of which would be quality air. Politicians would legislate from scientific facts not from their own personal and political welfare. Corporations would put human and planetary wellbeing above the profit stockholders investments. Profit would take second place to good global citizenship. Quality of life for all citizens of this planet would be our primary concern in our business and personal lives.

Whereas the efforts of one individual or those of many are to be encouraged and even lauded, we must not be lulled into thinking that individual actions alone will be enough. No, global warming is not going to lessen if we all use paper instead of plastic or drive electric cars, for example. Massive governmental and corporate actions and trillions of dollars will help turn the tide of global warming. Courageous leadership in the halls of governments and the boardrooms of corporations as well as on the streets and in the homes of the people of all nations is the only thing that will win this battle against climate change.

Don't get me wrong; individual actions are important and can make a significant difference, but individual actions alone, however, will not make this planet any healthier for us or for future generations. The global warming scourge requires actions across all segments of society. Nonetheless, here are some examples of what good stewards of this planet can do to reduce their carbon footprint:

* Drive a more fuel-efficient car.
* Combine trips. Don't make multiple trips to several different places.

* Walk or bike more places. It not only helps the environment; it is good for you as well.
* Use public transportation.
* Car pool.
* Drive a hybrid or electric car.
* Drive smarter: slower accelerating and driving.
* Keep your vehicle well tuned so it doesn't waste fuel.
* Use bio-fuel, e.g., mixture gas and ethanol, etc.
* Turn down your thermostat.
* Dress warmer when in your home.
* Make sure your home is well insulated and windows and doors don't let air in.
* Check to see if heating and air conditioning units are running efficiently.
* Keep a clean filter in your heating and air conditioning units.
* Use your air conditioner less and open your windows.
* If you must use your air conditioner, use it as little as possible and don't set it so low.
* Update your appliances to more efficient ones.
* When possible hang clothes out to dry instead of using dryer.
* Turn off lights and other appliances, e.g., TV radio, etc., in rooms when nobody is in room.
* Use LED bulbs.
* Eat less or no meat. Raising animals is environmentally costly.
* Avoid use of plastics. Whenever possible, use paper or glass.
* Buy in bulk whenever you can.
* Plant trees or other green plants.
* Use only dead wood for burning. Do not cut down live trees unless absolutely necessary.
* Get involved in environmental activities and projects.
* Join others doing the same.
* Support organizations advocating for a healthier environment.

* Talk to family and friends about your concerns for the planet and the next generation.
* **VOTE! VOTE! VOTE!**

Readers will notice a common theme in all the above suggestions, and that is sacrifice. Sacrifice is the act of giving up something important to oneself for the sake of another. Virtually all of us have made sacrifices for the good of others at least some time in our lives. James Traub's brilliant essay in the *New York Times* (2020), "Our 'Pursuit of Happiness' Is Killing the Planet" makes this cogent point: "Your freedom to live as you wish turns out to jeopardize public well-being." The time for sacrifice is now.

To address robustly the many problems of climate change and to reduce its impact on our lives, we must reduce our carbon footprint. This will require a level of sacrifice we are not accustomed to. The consensus of opinion among climate scientists is that our lifestyles are unsustainable. We cannot go on living as if we are the only ones that matter, as if global warming is not a real and present danger and as if our children do not deserve a more life-friendly Earth than the one we are leaving them.

If we continue our single-minded pursuit of personal pleasures, the Earth will continue to overheat. During an earlier time of crises, the Great Depression, FDR was able to call us to a higher cause, one greater than our individual pleasures. He asked and laws required all Americans to make sacrifices and they did. Such a national and global call for sacrifice is needed now. Who will step up and challenge us to such a sense of collective purpose again. Is there a leader in our Congress, our religious institutions, or our many international organizations who will ask us to make sacrifices again?

One final thought about sacrifice. Sacrifice must not be thought of as a negative, an unwanted action. In fact, I do not even like the word because it implies the loss of or the giving up of something that one does not really want to lose or give up. For the sake of clarity, I must speak personally here. My parents had eleven children. My father, a disabled veteran, died when I was three and my mother thirty-six. We were dirt poor before he died and dirt poor after. Realizing the seemingly impossible task of raising nine minor children (one brother had died at age four from diphtheria and one had finished high school early and joined the Navy so as to relieve some of the economic pressure on the family), my father's family urged my mother to give her children up to the children's home as it was known in those days (1944). My mother would not hear of that.

You have to understand my mother. She had lost every member of her family during the influenza outbreak of 1918, similar to the effects the novel coronavirus pandemic is having today. She was raised by a maternal aunt. She was determined her children would be raised and loved by their real parent not in an impersonal home by a stranger. My mother never complained about her decision or her situation though she would talk about the day her ship would come in. My mother never thought about her decision as a sacrifice. To her it was a calling, her duty, not a sacrifice. I consider my mother a saint.

To further illustrate the spiritual aspect of sacrifice, allow me to share a brief excerpt from the book, *A Traveler from Altruria*, by William Dean Howells. Mr. Homos, the traveler, is puzzled by the inequalities and social class distinctions he is observing in American culture. When asked how it is in his country, Altruria, the following dialogue occurred:

> "Why we have solved the problem in the only way, as you say, that it can be solved. We all live alike."

> "Isn't that a little, just a very trifling little bit monotonous?" Mrs. Makely asked, with a smile. "But there is everything, of course, in being used to it. To an unregenerate spirit-like mine, for example, it seems intolerable."

> "But why? When you were younger, before you were married, you all lived at home together. Or, perhaps, you were an only child?"

"Oh, no, indeed! There were ten of us."

"Then you all lived alike, and shared equally?"

"Yes, but we were a family."

"We do not conceive of the human race except as a family."
(italics mine)

Wow! What if we all considered each other family? Scaling back our standard of living would not be seen as a sacrifice. We would be willing to make changes in our lifestyles without feeling like we are sacrificing. Imagine such a world! By the way, where is this Altruria anyway?

Look again at the list above. Ask yourself what "sacrifices" you are willing to make so another generation may enjoy the blessings of this Earth that you have enjoyed? In his essay, Traub (cited earlier) asks a penetrating question: "Can we forge a new equilibrium before Miami is under water?"

Arguably, the most important action an individual can take is vote. That actually requires little sacrifice-- if one lives in the U.S., that is-- but it can make a huge difference. Politicians make the laws that affect you and the rest of the country. Their actions have great impact not only on you and the nation but also the world. To a large extent, politicians control, to a greater or lesser degree, the very quality of our lives and the lives of every person on this planet. Their action or inaction on global warming affects the whole of humanity. The very quality of our waters, soil and the air we breathe is determined by their actions.

In addition to the many attempts to roll back many of Obama-era environmental protections, one of the latest actions sanctioned by this administration is the misleading reports coming out of the Department of Interior headed by a Trump appointee. Specifically, as reported by Tabuchi (2020), an Interior Department employee, Indur Goklany, has embarked on a campaign to mislead the public about climate change. He has claimed in Interior Department reports that there is a lack of consensus among climate scientists that the Earth is getting hotter, that climate scientists may be overestimating the rate of global warming and that the increase in CO_2 is actually a beneficial thing! More disinformation from a fossil fuel partisan!

He is doing what deniers have done ever since it became clear that the Earth was getting hotter, that the increase in temperature was a threat to our health and wellbeing and that it is caused by human activity. He has taken a fact, the consensus among scientists about global warming is *only* in the range of 97%, and manipulated it to the administration's advantage. In addition, whereas scientists are not exactly certain about the rate of global warming, especially since they have discovered that more methane is being released than they originally thought, they agree that, in fact, we may be under-estimating how quickly the Earth will become hotter. And, finally, to his third assertion that increased levels of CO_2 are a beneficial thing, I guess he didn't read the research studies of scientists working for the United States Department of Agriculture. They have discovered that rice, upon which millions of people depend for their daily calories, actually becomes less nutritious as CO_2 levels increase. His whole assertion rests on the belief that if a little CO_2 is good for plants, then more would be even better.

Believe me, I have tried not to be overly critical of this administration, but, folks, we must face the fact that the policies and actions of the Trump administration are not going to protect our children and their children from the ravaging effects of climate change. As Margaret Mead once said to me following a lecture she had just delivered: "We get the politicians (speaking of Nixon) we deserve." I want to argue that point, but I really cannot. We must vote and vote for candidates, be they Republican or Democrat, who accept climate change and feel earnest about it. Our children need voters who understand the challenges of climate change and who are willing to vote informed by facts not ideology or political party.

Another thing we all must do, talk *with* our family and friends about global warming. Survey (see preface) data shows that most of us (63%) rarely or never discuss this topic with our family and friends. It is because people hold such strong opinions on this matter which makes it so difficult to have a calm, rational discussion about the facts. The reasons for these strong opinions were discussed in Chapters 3 and 4. Below are some suggestions to help you have this difficult conversation:

1. Notice the use of the words "talk with" rather than "talk to." "Talk with" assumes a dialogue whereas "talk to" assumes a monologue. Monologues do not show sufficient respect and respect is something people demand.
2. Never start the conversation until you have express permission to do so. Any attempt to impose your beliefs, even ones based on facts, on someone will be met with resistance. Imposition of your beliefs will invite argument because it is natural for people to defend their position when attacked. The more timid may not argue with you, but they will withdraw and not really listen. If you want to engage people in a conversation about global warming, try to find a natural way to bring the subject up as when, for example, there has been some recent news about it. Even then, get their permission to discuss the topic before you express your concerns. Then and only then may they be willing to engage in *dialogue*.
3. Once a dialogue starts listen more than you speak. Resist being a "teller." Listen and ask questions rather than preach or teach.
4. Help them see the difference between beliefs and facts. Gently! At the same time, respect their beliefs as you attempt to make this important distinction. You can point out that beliefs may lead us away from truth as in the case, for example, of "Flat Earthers" and even vaccination deniers who put their children at risk based on bad science. Be careful here. Know who you are talking with.
5. If they have a strong opinion, inquire about how and where they got that opinion. Try to understand why they believe what they believe. Above all, don't come across as judgmental or as a "know it all." Help them understand not all sources or beliefs are equal. Beliefs not based of facts can be as dangerous as we have

discovered with the Iraq War and Vietnam where we all were mislead about the threat. Again, you must be careful here too.
6. Be prepared to offer your facts and the source of your facts. Encourage them to read more about the subject at government websites starting with NASA and NOAA as well as university websites where much research is being done.
7. Have no rigid expectations about how the person "should" respond to you. Understand that some people will not care about "your facts." Beliefs make people comfortable and it's natural to resist beliefs or facts whose acceptance might cause them some degree of emotional turmoil. Understand some will want to remain in what you and would call their ignorance.
8. Be clear about the difference between respecting people and respecting all their beliefs. Never show disrespect for them as people. I am assuming your relationship with them does not depend on them accepting your position on global warming.
9. When all else fails, leave them with this puzzling question: 'Do you believe in the moon?" Sometimes this might even be a good way to start the conversation because the lesson is obvious. People believe in the moon because they can see it. Scientists (and you) believe in global warming because they and you can see the evidence.

Now, obviously some of us are more responsible than others because of the wealth of the nations in which we live. Those in countries with higher GNPs carry more of a burden because we are consuming more goods and resources making our carbon footprint greater. Like a good marriage where everything is not always equal, sacrifice will not always be equal either. Those who have more will have to sacrifice more. That is the only way we will see any reduction in our carbon footprint. We can't wait for all to sacrifice equally. Someone must lead the way and bear the heavier burden perhaps--just like it is from time to time in a marriage.

The challenge of any generations is to protect the planet for future generations. Previous generations were ignorant of the problems of climate change. We can excuse their inaction, but you and I have no such excuse. We can no longer ignore the problems of climate change. We cannot! We are the only species on this planet that has a choice about how we treat this Earth and each other. Only humans hold the destiny of this planet and future generations in their hands. It is time we live not just as smart *Homo sapiens* but as responsible ones as well.

Climate change demands immediate action. We cannot allow our governments to delay taking effective action to protect our environment from ourselves as the US government has done, for example, with social security and the national debt. This is not another issue that can be kicked down the road, left for another generation to clean up our mess. Unless we face climate change as fundamentally moral and spiritual issues and not a political one, we will condemn future generations to an Earth on which life is unimaginably difficult for all and even impossible for many, if not most.

Actions at all levels of society will be required to stem the tide of global warming. If we all put aside our own self-interests, focus on what is good for the Earth and the common good, and cooperate with each other with a commitment to save the planet for future generations, we will reduce the harm associated with global warming for ourselves and future generations. If we do not, well then, someone will pay the price for our neglect and abuse.

In our efforts to influence public opinion, we must remember the role of brand connections (see chapter 4). We must appeal to people's unconscious processes as successful marketing people do. We all have seen countless ads pairing a product with a slim, sexy female. Such an association sells products. How much more successful would we be at influencing public opinion if we borrowed ideas from the advertising industry? What if the facts of global warming and actions to reduce it, were presented by a rock or movie star? What if there was a concerted campaign to associate good Earth stewardship with good citizenship or admirable behavior? What if efforts to protect our planet and humanity from the ravages of climate change were seen as "cool?"

Make no mistake about it, this will be a long fight. If human history has taught us anything, it is that power does not yield to reason; power can only be overcome by power. The fossil fuel industry will not yield its power until it is forced to do so. Politicians who serve their own self and political interests will continue to do so until the power shifts from those who deny or ignore anthropogenic climate change to those who love the Earth.

The power of reason is limited. Reason did not keep us from invading Iraq during the George W. Bush presidency. The lessons of Vietnam were lost on his presidency and the American public who backed our war with Iraq. Reason has not stopped genocides from occurring since the Nazi Holocaust. Reason does not prevail in our national elections. Reason does not stop us from making bad purchases. Reason does not keep people from committing another crime that is sure to land them back behind bars. For the sake of our survival, we were hard-wired to react emotionally first, but we must overcome this survival response and think more rationally if we are to solve the problem of global warming.

Yet, I am not ready to stop appealing to reason because I still believe reason can shift the balance of power. We are, after all, still rational animals for the most part. This can be seen in the fact that many are waking up to the dangers of global warming and are beginning to demand change. Reason is having that effect. Survey data shows that each year more and more people are taking global warming seriously and are beginning to demand our governments do something about it.

Jon Meacham in his book (2018) offers actions each citizen must take to save our democratic way of life. These actions are the same actions all citizens of this planet must take to save our Earth:

1. Enter the arena. He writes, "The battle begins with political engagement itself." He quotes Theodore Roosevelt: "The first duty of an American citizen, then, is that he shall work in politics; his second duty is that he shall do that work in a practical manner; and his third is that it shall be done in accord with the highest principles of honor and justice." Working to save our planet from global warming is noble and just work. Many of us want nothing

to do with politics. We must remind ourselves, however, that it is only in the political arena where the change that will save our planet will come.
2. Resist tribalism. I would add resist making global warming a partisan political issue. Listen to reason. Listen to those with opposing points of view. As my mother would say, there are many ways to skin a cat. All ideas about controlling global warming should be considered. One should not, however, compromise the noble and just work of combating climate change for the sake of agreement.
3. Respect facts and deploy reason. Know the science and let the science guide our actions. Too much is at stake, as we have discovered during the coronavirus pandemic, to listen to people and politicians who ignore the science.
4. Find a critical balance. Meacham writes: Being informed ..."entails being humble enough to recognize that only on the rarest occasions does any single camp have a monopoly on virtue or on reason." We should listen to the fossil fuel industry. After all, they have a place at the table as we wrestle with how to preserve our planet and a reasonable way of life for all Earth's inhabitants. We need to quit seeing them only as the enemy and invite them to join our efforts at combating climate change. Perhaps this seems naïve, but until we try it we will never know.
5. Keep history in mind. The battle over global warming is a history of lies and misinformation on the one hand and excessive claims of destruction on the other. A knowledge of our history in this battle would serve us well as we learn to work together to save our planet.

Throughout our history, women and mothers have made a profound difference in public attitudes and policies. It was, for example, women who forced our government to pass the 19th Amendment allowing women the right to vote. A grieving mother formed an organization to fight drunk driving, MADD, Mothers Against Drunk Driving.

These are but two examples of women causing important social change. Consider, therefore, the words of Mrs. Carroll Miller of Pennsylvania. Mrs. Miller was speaking at the 1924 Democratic Convention in favor of a plank in the platform condemning the KKK, Ku Klux Klan, when she challenged the few newly enfranchised women in attendance to support the plank. She said: "If men are afraid to face this issue, I beg you cast aside your trepidation and deceit. We who are accustomed to suffer the pains of childbirth that the race may go on should not be afraid to uphold a great principle that our children may live in happiness and security. We who are accustomed to wait and fight in the lonely watches of the night for the life of the child when death is hovering over the crib should not be the ones who flinch now."

I am asking women to take on the critical issue of climate change for the sake of their children and grandchildren. Climate change is this generation of women's great principle affecting their children's happiness and security. There are certainly some good men fighting the fight, but a strong coalition of women could tip the scales in favor of effective actions to slow down global warming. Such a coalition is the missing ingredient now.

We all would do well to remember the words of the late civil rights icon and Congressman John Lewis: "My philosophy is very simple. When you see something that's not right, not fair, not just, you have a moral obligation to do something, to say something, to speak out." Global warming is one of those not right, not fair, not just things. It may not be affecting you personally or directly, but it will your children and their children for generations to come.

"The times they are a changing" sang Bob Dylan years ago. We see that in so many ways. For example, Greta Thunberg has awakened the soul of many people. Her school boycott was the beginning. Her courage to speak so forcefully and directly at the United Nations emboldened others to act courageously. Lovers of this planet like Greta, Rep. Alexandria Ocasio-Cortez, and Jamie Margolin have sparked a revolution addressing the problems of climate change. Groups like Sunrise Movement, Extinction Rebellion, Zero Hour, National Children's

Campaign, Schools for Climate Action, and 350.org have organized to combat climate change. They are all trying to shift the balance of power to favor legislation and practices that protect our planet. The combined voices and power of each lover of the Earth will be required to force corporations and governments to create policies and practices that favor the Earth and future generations. Respond to a call to arms! Join the fight to save our planet for future generations. Please! We *can* save this planet but the real question is *will* we?

Discussion Questions

1. What will it take to stop or diminish global warming/climate change?

2. Are you more like Ms. Marris who sees the glass as half full or more like Ms. Kolbert who sees it as half empty in respect to addressing the problem of climate change?

3. Like Ms. Kolbert do you see no end to our incessant demands for more? Is this incessant demand for more a part of the global warming problem? If so, in what ways?

4. What is the meaning of the author's statement that prevention begins with realism? Are we as a people and our government being realistic in addressing the problem of climate change?

5. Would you be willing to pay more for your electricity if it was produced by green technology, for example, wind, solar, tidal or geothermal? How much more each month? Check your state for price comparisons.

6. From the list of actions that would reduce global warming, what three are you willing to adopt? What three would you find most

difficult to do? What three would your children want you to choose? Would it seem like a sacrifice?

7. The excerpt from "A Traveler from Altruria" speaks of the whole human race as one family. Do you see the whole human race as your family? If so, how does that affect your actions on this planet? If not, same question.

8. Are you comfortable enough with the issue of global warming/climate change to discuss it with family members and/or friends? Do you think you could have a productive, agreeable discussion with others? Why or why not?

9. Discuss Meacham's recommendations as they apply to global warming.

10. Can a coalition of dedicated women make a real difference in the climate change debate?

9

The Beginning or the End?

"We're just speaking to every human soul, to every human brain, to every human heart to say, 'Who are you? What do you want to do on this planet? How do you want to be remembered?"

-Christiana Figueres, former Secretary of the UN Framework Convention on Climate Change

"If we continue to accumulate only power and not wisdom, we will surely destroy ourselves."

-Carl Sagan, astronomer

Dr. Elena Irwin, Faculty Director of Ohio State University's Sustainable Institute, offered three scenarios regarding our future here on this planet while speaking online at OSU's 50 year celebration of Earth Day. Which world we create depends on what policies we implement now. A condensed version of her three scenarios is offered below:

World #1: Business As Usual where we "sleep walk" our way into the future. We fail to meet the goals set by the IPCC of 45% global net reduction in CO_2 from 2010 levels by 2030 and fail to meet the goal of zero emissions by 2050. Because of the absence of a world-coordinated approach, masks have to be worn when we venture outside as the air is so polluted. Places on Earth have become uninhabitable because some places are too hot while others have been inundated by rising seas. Environmental changes have forced mass migration like we have never seen before creating great social and political instability. New diseases have emerged in our new environment. Economic damage is widespread throughout societies.

World #2: Global Techno Economy where we have put our faith in competitive markets and innovations and the motto of our time is "There are no jobs and no customers on a dead planet." We continued to rely on fossil fuels and there has been much economic growth. People are better off economically. Education and health care are better around the world. Even though we surpassed the IPCC goal of cutting our net emissions in half by 2030, we did meet the goal of net zero emissions by 2050 through renewables and massive scaling of negative emission technology. Coral reefs collapsed and the Arctic sea is now passable during the summer. Geo-engineering helped us avoid a worst-case scenario, but there is concern that we will not be able to keep the Earth from warming beyond 1.5 degrees F.

World #3: Ecological Balance is a world that woke up due to the novel coronavirus pandemic of 2020. We realized the seriousness of the problem. The world came together. Governments and corporations worked cooperatively to solve the problem of global warming. We met all our IPCC targets and fully decarbonized our electricity output. We are on a safe path now. Air is slightly more breathable and more people have access to clean water. We took a "safety first" approach. Not all people prosper. Poor, undeveloped countries and their people still struggle. Social inequities remain and we struggle to create a shared prosperity for all.

Dr. Irwin's analysis is sobering. Global warming represents a "catch 22" scenario where we are damned if we do and damned if we don't. With our initial and necessary reliance on fossil fuels to run our economies, we have created a situation where there are no easy solutions. We did not know any better when we first started to burn fossil fuels. Only recently have we realized what we have done to our Earth. Now, if we ignore the problem, it clearly will be to our own peril-world #1. If we act too robustly, however, we create more social and economic disparity-world #3. A moderate approach, world #2, will improve our health and the health of our planet in the long run, but at cost to our environment as well as ourselves in the short run. We have created a Pandora's box.

A minister of God, Paul K. Chappell, wrote (2009): "Although I have learned that war is not inevitable, I have realized that a peaceful future is also not inevitable. War will not magically end on its own, and our world will never know peace if we sit around and do nothing. War will only end if we end it. War will only end if we stand up and make a difference, if we take steps to ensure our survival and prosperity as a global community."

We must decide which world we want to create. Chappell's point is clear and it applies to global warming as well as it does to peace. An effective solution to global warming is not inevitable. Global warming will only end when we end it. Global warming will only end when we all take a stand against harming our planet further. To end global warming, we must first stop the denial. We must take global warming more seriously like we are finally taking the coronavirus seriously. We must recognize that we are not invincible or that we have plenty of time to address the problem of global warming. We must recognize when "solution aversion" (chapter four) is in play.

Second, we must put our love for our Mother Earth and for all humanity above our own narrow self-interests. We must be willing to deny ourselves some of the pleasures and conveniences we all are addicted to. We must quit thinking "we can have our cake and eat it too." We must not only change our individual habits and lifestyles, we must insist our governments support the transition to green energy and make the necessary changes in our economy to support a healthy Earth. When we have taken such steps, we will have made progress to ending global warming and restoring health to our fragile planet.

Robert K. Watson, founder of LEED, Leadership In Energy and Environmental Design, offered this morsel for our consideration: "Mother Nature always bats last, and she always bats 1000." As the old saying goes, "don't mess with Mother Nature." In human relations we know that what we send out we get back. The same is true of Mother Earth; what we send out (into our atmosphere), we get back (pollution and overheating).

Thomas Friedman (2020) identified two rules for climate change mitigation as well as for mitigation of a pandemic. The first is to manage the unavoidable so you can avoid the unmanageable. The second is if we don't take care of the conditions that enable us to prosper, we will not prosper in the end." The unmanageable can still be avoided. Our climate dilemma is manageable if we act now. We can continue to prosper if we act smartly in respect to both our economy and our environment. We violate these rules at our own peril.

America, the world, we are at a crossroads. Either we take global warming, climate change, seriously or we do not. Time for delay is over. Time for denial is over too. We must act now and we must act with haste, but wisely. As I write we are witnessing the harmful effects of waiting too long before we act. One thing the coronavirus should have taught us is that delay in responding only makes matters worse. We must abandon false security measures and face the problem head on. Not tomorrow, not next month, not next year, but now, today.

In all fairness, many have heeded the call of Mother Nature and suffering humanity and have started taking steps to reduce our carbon footprint. Individuals, businesses, corporations, and governmental bodies of all sizes have begun to act to prevent further warming of our planet. Much is already being done, but, oh, so much more needs to be done.

It is fair to say we will not likely reach the original 2015 Paris Agreement goal of limiting Earth's heating to 1.5 degrees C, especially if we continue on the course we are on. There simply has not been the level of international support to make that happen. Too many countries have just not taken the problem seriously enough and some countries are not advanced enough to commit to such a noble goal. Some that can begin a robust effort, such as the United States (Trump) and Russia (Putin), are still resisting the science and the necessary transition to alternative energy. Countries like India and China are trying to advance their global position by producing more and more goods, relying on fossil fuels to do so. At the same time though, China is increasing its use of electric trains and buses to reduce pollution.

So, here we stand on the threshold of tomorrow's Earth deciding whether we alive today want to make the "sacrifices" necessary to secure a healthy planet for our children and their children. Never before have we had to face such a challenge. Oh, sure, we have had challenges we have had to face and we have done a yeoman's job of doing so--diseases for which there was once no cure or treatment, rural electrification, sanitation issues, interstate transportation, global transportation, almost instant world-wide communication, and food production and distribution, for example. When we put our collective minds to it, there is little we cannot do. That is, when we put our collective minds to it!

We are at the threshold of a new Earth, a new humanity. Global warming is testing whether we can learn to live together cooperatively and in harmony. We are being tested to the depth of our morality and spirituality. To borrow from Lincoln's address at Gettysburg at a critical moment in our Civil War, global warming is testing whether nations of the world, humanity, can long endure and continue to prosper. Referring to that great president one more time, "A house divided against itself cannot stand." We are discovering that an "Earth" divided against itself cannot stand either; it cannot prosper. Justice cannot prevail and peace cannot be secured until we learn at long last to live together respecting the rights and needs of each other and of each nation.

We need Scott Kelly's "space perspective," (2020): "Seen from space, the Earth has no borders. The spread of the coronavirus is showing us that what we share is much more powerful than what keeps us apart, for better or for worse. All people are inescapably interconnected, and the more we can come together to solve our problems, the better off we will all be." We must give serious consideration to his perspective and that of every astronaut who has had the opportunity to see the Earth from the advantage point of space. We are intricately connected and we must move forward with that perspective if we are to solve the problem of global warming.

Seven million years of *Homo sapiens'* existence has come down to this momentous point in human and planetary history. Will we continue to prosper or will we, because of our abuse and neglect of our Mother, condemn the Earth and its inhabitants to a misery we have never known? The choice is ours. Unless there emerges a willingness to cooperate and sacrifice as a humanity, as nations, for a greater cause, our "house will not stand." As Carl Sagan suggested, unless we get smarter about our life here on this planet, we will destroy ourselves. We will choke and burn in our own foul and hot environment.

If we can achieve environmental justice, we can prosper and survive. If, in other words, we can fairly treat and involve all people regardless of race, color, national origin, or income in the development, implementation, and enforcement of environmental laws, regulations and policies, we can create a healthy environment for all humanity. Even if you are not a person of faith, and especially if you are, we all should pray for cooperation and sacrifice like humankind has never witnessed before. Either we learn to flourish together or we will decline together.

Collectively, we cannot create heaven, but collectively we can save ourselves from hell. All it takes is summarized by the prophet Micah (6:8): "Act justly, love mercy, and walk humbly." To walk kindly upon the Earth among all her inhabitants, all of humanity, without regard to anyone's national origin and with love of our Earth, will save us from ourselves and our foolishness. We need a Micah now to lead the way toward a more life-friendly Earth for all humanity. We all can be a Micah. Will you be one?

The noted poet, John Keats, offered us a perspective we would do well to heed when he wrote about "negative capability." He defined it as the state of "being in uncertainties, Mystery, doubts, without any irritable reaching after facts and reason." In other words, he counsels us to embrace reality, accept the facts and act with reason. For far too long many of us and our leaders have been doing just the opposite. Humanity cannot afford for us to continue to turn a blind eye to the existential problem of global warming and climate change. We do so not only at our own peril but also the peril of every generation who follows us.

I want to close this book on a positive note. After all my research and thoughtful consideration of all that I have read and written, I still believe in humanity's collective wisdom. We will solve the problem of global warming and its resultant climate change. We will. I suspect it will be a few more years before governments become really serious about it and act accordingly, but we will. Right now all of us are more concerned about the COVID 19 pandemic, as we should be. That pandemic is the most immediate threat to our health and wellbeing, to life as we now know it. Once we are through that crisis, we will turn our attention to the "pandemic" of global warming. We will have learned the lesson that delaying the inevitable only makes matters worse.

We will see through the attempts of the fossil fuel industry to delay or prevent actions on climate change. We will accept the scientific facts about global warming and climate change. Pressure will mount on politicians who serve as fossil fuel agents in Congress to start earnestly addressing the problem of global warming. A new president will be elected who believes in anthropogenic global warming and will lead the country and the world into a new direction, one that respects life and this planet above profits, Dow Jones and GNP. People will be willing to accept changes in their life styles for the sake of suffering humanity around the globe and for the sake of generations to come.

Some day soon, within the next decade if not sooner, the realities of global warming will be impossible to ignore. If we haven't awakened to the realities of global warming before then, we will be forced to. We will not leave our children and grandchildren the planet we have enjoyed; it is already too late for that. We will, however, make great strides in stopping the ruin of this planet, our Mother Earth for them. We will! We can do it. As our mothers always said, "Were there is a will, there is a way!"

We would be wise to consider the words of "the Old Boy," Lao Tzu, founder of Taoism, author of *Tao Te Ching* or *The Way and Its Power*, a slim volume of 5000 Chinese characters. According to one legend, Lao Tzu was saddened by man's disinterest in cultivating the natural goodness he advocated. Realizing he could do little to change man's disinclination toward the more sublime, one day he climbed on the back of a water buffalo and headed west toward what is now Tibet. According to a legend written about by Huston Smith, *The Religions of Man*, at the Hankao Pass the gatekeeper tried to persuade him to turn back. When it was clear Lao Tzu was not going to be convinced to stay, the gatekeeper prevailed upon him to leave a record of his teachings to those he was leaving. Today his wisdom is known by millions around the world.

One of his pearls of wisdom relates to our treatment of the Earth. He wrote:

> *"Those who would take over the Earth*
> *And shape it to their will*
> *Never I notice, succeed.*
> *The Earth is like a vessel so sacred*
> *That at the mere approach of the profane*
> *It is marred*
> *And when they reach out their fingers it is gone."*

Amazingly, this is the perspective of a man who lived some 2,600 years ago, before large-scale mining of the Earth's precious resources and the burning of fossil fuels! He warned that trying to control the Earth for our own gain would forever change the Earth resulting in a loss of the Earth we know. Global warming is evidence of this profanity is it not? We are in danger of losing the Earth as we have known it. Already changes in our atmosphere and on the land are abundantly obvious. Unless we heed his wisdom we are in danger of losing the Earth that has sustained us so adequately since our beginnings some 250,000 years ago.

Discussion Questions

1. How do you answer the three questions posed by Christiana Figueres at the head of the chapter?

2. Of the three worlds presented by Dr. Elna Irwin, which do you prefer for yourself and your children?

3. Does our future feel like a "Catch 22" to you?

4. What does the author mean when he suggests that "we want our cake and eat it too?" Is it possible to have our cake and eat it too?

5. What does Robert Watson mean when he says Mother Nature always bats last, and she bats 1000?

6. In reference to what Thomas Friedman says, are we managing the unavoidable so we can avoid the unmanageable? How well or how poorly are we doing?

7. The author writes, "Global warming is testing whether we can learn to live together cooperatively and in harmony." How are we doing?

8. How hopeful are you that we will save the planet for our children and their children?

9. Is there more you can and should be doing? Your local, state, and federal government? Nations of the world?

10

Hopefully

"Hope and fear cannot occupy the same space at the same time. Invite one to stay."

-Maya Angelou

I feel compelled to make one final appeal for our Mother Earth, for humanity and for all forms of life on this wonderful planet. Hopefully, after reading this book and perhaps after exploring the topic even further, you are convinced that life as we know it is being seriously threatened by global warming and climate change. Hopefully, you have no doubt in your mind that it is caused by human activity. Hopefully, you now understand global warming is not a hoax, not a plot to ruin capitalism and institutionalize socialism, install a world government, make huge profits, advance a liberal agenda, or whatever other conspiracy theory you may have heard about. Hopefully, you understand that the sooner we act, the better off all of us will be.

Hopefully, we will learn lessons from the coronavirus pandemic that to deny or minimize a problem only creates more suffering in the long run, that the best time to solve a problem is earlier rather than later before the problem overwhelms our ability to respond, that some natural problems are not easily fixed once the problem is allowed to worsen, that some problems require world-wide cooperation and a coordinated response, that prediction models are not always perfect but yet essential for planning, that the less fortunate are at greater risk and will suffer more than those more fortunate and that the solution to some problems require great "sacrifice" on the part of us all especially if we are slow in responding.

Hopefully, you care about your Mother Earth and all its inhabitants enough to make "sacrifices' to preserve it for future generations. Hopefully, you love them enough to look beyond your own wants and needs and consider what is best for others, including those not yet born. Hopefully, you want for your children and grandchildren an Earth like so many of us has enjoyed. Hopefully, Abraham's children and all Earth's inhabitants will rise up and demand better from our leaders and our politicians. Hopefully, all nations of this beautiful planet will learn to cooperate to stem the tide of global warming. Hopefully, someone will emerge who can unite all of us, all nations in this global fight for the Earth our children and grandchildren deserve.

We are a young species on this planet; yet, we are the only species who can save this planet from ourselves. We can make or break Earth as a hospitable planet. Mother Earth has always wanted us to prosper. She has steadfastly provided for our needs. She has indulged us like an over-indulging parent. But, now she needs our help. Her health is bad and getting worse each year. Hopefully, we love her enough to take care of her in her time of need.

Walter Brueggemann (2018) had this to say about hope: "Hope, on one hand, is an absurdity too embarrassing to speak about, for it flies in the face of all those claims we have been told are facts. Hope is the refusal to accept the reading of reality which is the majority opinion; and one does that only at great political and existential risk. On the other hand, hope is subversive, for it limits the grandiose pretension of the present, daring to announce that the present to which we have all made commitments is now called into question." We are at a juncture in history where we must call into question how we have been treating our planet and each other.

Hopefully, our children will look back across generations and be able to appreciate the Earth we left for them. Hopefully we will understand what Viktor Frankl meant when he said (*Man's Search for Meaning*), "In some ways suffering ceases to be suffering the moment it finds a meaning." I can think of no greater meaning than for all citizens of the Earth to dedicate ourselves to solving the problem of global warming and climate change. Let's each do our part. Let's get it done!

In conclusion, I am hopeful about our planet and our people to the extent we learn the invaluable lessons from the coronavirus. There are at least three lessons we must learn from this pandemic: the importance of effective leadership, the value of science in forming policies, and the necessity for community cooperation.

If the coronavirus has taught us anything, it has taught us the critical value of effective leadership. A successful plan to address global warming/climate change demands a leader who will face the reality of the problem. Such a leader will be more concerned about the welfare of the people more than he (she) is his own personal and political fortunes. This leader will pay whatever price personally and professionally to address the problem effectively. He will have a strategic plan than finds the right balance between the economic interests of his constituents and his nation and the people's health and wellbeing.

This leader will be creative in finding ways to fund an effective strategic plan that addresses climate change. He will spare no expense in combating climate change because he recognizes the gravity of our situation, the real threat of continued global warming. He will have the will and courage to make tough and probably unpopular decisions. His commitment will always be to the health and wellbeing of the people not his own self interests and that of his political party.

Secondly, the leader will also believe the science about climate change. His constituents will follow his lead in the acceptance of science. Science will be the basis on which he and other policy makers will make policy. He will not attempt to distort or deny the science or to silence our scientists. Instead, he will challenge our scientists to produce their best results and to perfect their climate models. He will not listen to those who tell him what is in his best political and personal interests, especially when the health of the planet and its inhabitants are at stake. Instead, he will listen to the scientists who tell him what policies protect the interest of the health and wellbeing of the people and this planet.

Lastly, an effective leader will be successful in securing the commitment of the public to his clearly communicated goals for our lives and our planet. He will be able to convince people of the necessity of their cooperation in a well formulated, bipartisan plan to fight global warming. He will help them see the importance of making personal "sacrifices" for the sake of generations to come. People will know that he has their best interest at heart. He will model the behavior he wants the public to engage in and will also personally make sacrifices necessary to preserve a healthy planet for future generations. He will win their support for effective climate change policies. He will not waiver in the face of public pressure or negative press. Instead, he will lead us along a path to restoring the health of our people and our planet.

Hopefully such a leader will emerge. Our future and the future of this wonderful, magnificent planet depend on what we do now. Scientists advise us that now is the time. We dare not delay any longer in attacking climate change. Our fate is in our own hands. May Homo sapiens rise to the challenge and earn their *"sapien"* name.

Discussion Questions

1. How hopeful are you about the future of our planet? Will we be passing onto our children a planet that will be favorable or unfavorable to them?

2. What part will you play in the future of our planet?

3. Will your children and grandchildren praise or condemn you as they inherit this new Earth we are leaving them?

4. Is hope a dangerous thing?

5. Are you willing to "suffer" some so that future generations will inherit a better Earth?

6. What changes are you willing to make?

7. The author listed five objectives he hoped his book would accomplish (pp. xxix-xxxi). Was he successful in accomplishing them?

Appendix A

Spiritual Health Inventory

Directions: This is not a test where there are right and wrong answers. These statements are to be used to facilitate introspection and discussion about spirituality not as a measure of it. Do not be concerned about what the "right" answer might be or about what others might think of you-if you decide to share your responses. Indicate to what extent a statement is either true or not true of you and to what degree by selecting one of the following responses and placing the letters of that response in the space provided: **VL**: very much like me; **SL:** Somewhat like me;

LL: Little like me; **NL**: Not much like me; **DL**: Definitely not like me.

1. _____ Throughout the day there are many times I think of how blessed I am.
2. _____ There are some people I just cannot forgive.
3. _____ I sometimes think about how much suffering there is in the world and feel a sense of sadness about it.
4. _____ I make sure I have some period of quiet and solitude each day during which I may pray, meditate or just become immersed in the silence.
5. _____ Even when I know I am wrong, sometimes I have a hard time admitting it or agreeing with the other person.
6. _____ There have been moments when I have felt a sense of oneness or connection with God and/or the whole of humanity.

7. _____ Sometimes I actually feel some guilt about how good I have it compared to the plight of so many others.
8. _____ I feel a sense of serenity about myself and my life.
9. _____ I tend to be friendly and respectful with store clerks and strangers, treating them like they are important too.
10. _____ Even when I don't agree with people I try to understand their point of view instead of automatically trying to convince them of my point of view.
11. _____ I find myself at times wishing misfortune on someone I don't like or who has offended me.
12. _____ Though I realize anything could happen to anyone at anytime, I am optimistic about my future, especially my ability to cope with whatever life sends me.
13. _____ I often read materials that inspire me to be a better person, materials that promote my spiritual growth.
14. _____ I find myself becoming impatient with people who don't agree with me or who are annoying.
15. _____ I believe the universe is unfolding as it "should" and that humanity as a whole is evolving to higher levels of consciousness or spirituality.

Trusted Resources*

Climate.gov

Discover Magazine (Science for the Curious) *@discovermagazine.com*

George Mason University for Climate Change Communication

@climatechangecommunication.org

Intergovernmental Panel on Climate Change @ ipcc.ch

Mann, Michael E., *The Hockey Stick and the Climate Wars*

Mooney, Chris, *The Republican Brain: The Science of Why They Deny Science-and Reality*

National Aeronautics and Space Administration @ *nasa.gov*

National Climate Assessment @ globalchange.gov

National Geographic (especially 04/2020) *@nationalgeographic.com*

National Oceanic and Atmospheric Administration @ *noaa.gov*

Primack, Joel R. and Abrams, Nancy Ellen, *The View from the Center of the Universe*

Scientific American (especially *Special Collector's Edition*, Winter 2017/2018)

Sukys, Paul, *Lifting the Scientific Veil: Science Appreciation for the Nonscientists*

350.org

The Atlantic

The Water Will Come by Jeff Goodell

The Uninhabitable Earth by David Wallace-Wells

Union of Concerned Scientists *@ucsusa.org*

Union of Concerned Scientists, *Cooler Smarter: Practical Steps for Low-Carbon Living*

The New York Times, (source of many science articles)

Tyson, Neil deGrasse, *Astrophysics for People in a Hurry*

World Meteorological Organization @wmo.int

Yale Program on Climate Change Communication *@climatecommunication.yale.edu*

Youth to Power: Your Voice and How to Use It, Jamie Margolin (a handbook for youth)

*These are not the only sources to be trusted; however, if you find information that runs counter to what these sources report, it should be questioned.

Bibliography

Preface

D'Angelo, Chris and Kaufman, Alexander C. "Pentagon Confirms Climate Change Is a National Security Threat." *Huffington Post,* January, 18, 2019.

Spong, John Shelby. *Here I Stand: My Struggle for a Christianity of Integrity, Love & Equality.* New York: HarperCollins Publishers Inc. 2000.

Cooper, Anderson. *People Magazine.* June 22, 2020.

Chapter 1

Hansen, James. "The Greenhouse Effect: Impact on Current Global Temperature and Regional Heat Waves." *US Senate Committee on Energy and Natural Resources, Hearing.* June 23, 1988.

Bowen, Mark. *Censoring Science: Inside the Political Attack on Dr. James Hansen and the Truth of Global Warming.* Penguin Group. 2008.

Michaels, David. *Doubt Is Their Product: How Industry's Assault on Science Threatens Your Health.* Oxford University Press. 2008.

Oreskes, Naomi and Conway, Eric M. Merchants of Doubt: *How a Handful of Scientists Obscured the Truth o Issues from Tobacco Smoke to Climate Change.* Bloomsbury Press. 2010.

Westervelt, Amy. "How the Fossil Fuel Industry Got the Media to Think Climate Change Was Debatable." *Washington Post,* January 10, 2019.

Popovich, Nadja, et.al. "The Trump Administration Is Reversing 100 Environmental Rules Here's the Full List." *New York Times,* December 21, 2019.

Union of Concerned Scientists. "The Climate Deception Dossiers." June 29, 2015.

Broder, John. "Climate Change Doubt Is Tea Party Article of Faith." *New York Times,* October,20, 2010.

Mann, Michael E. *The Hockey Stick and the Climate Wars: Dispatches from the Front Lines.* Columbia University Press. 2012.

Seitz, Frederick. *The Wall Street Journal.* June 12, 1996.

Krugman, Paul. *Arguing with Zombies: Economics, Politics, and the Fight for a Better Future.* W.W. Norton Company. 2020.

Hegghammer, Thomas. *Jihadi Culture: The Art and Social Practice of Militant Islamist.* Cambridge University Press. 2017.

Ropeik, David. *How Risky Is It Really: Why Our Fears Don't Always Match the Facts.* McGraw- Hill Company. 2010.

Chapter 2

Spong, John Shelby. *Here I Stand: My Struggle for a Christianity of Integrity, Love &Equality.* New York. HarperCollins Publishers, Inc. 2000.

Butler, Octavia. "A Few Rules Trying to Predict the Future." *Essence Communications, Inc.* 2000.

Wayne, Graham P. The Beginner's Guide to Representative Concentration Pathways. *Skeptical Science,* Version 1.0. August 2013.

Miller, Kenneth. "Your Brain on Tech." *Discover,* May, 2020.

Chapter 3

Zuboff, Shoshana. *The Age of Surveillance Capitalism.* Perseus Book Group. 2019.

Chapter 4

Campbell, T.H. and Kay, A.C. "Solution Aversion: On the Relationship between Ideology and Motivated Disbelief." *Journal of Personality and Social Psychology,* 107 (5), 809-824. 2014.

Mooney, Chris. *The Republican Brain: The Science of Why They Deny Science--and Reality.* John Wiley &Sons, Inc. 2012.

Edsall, Thomas B. "The Whole of Liberal Democracy Is In Great Danger at This Moment." *New York Times.* July 22, 2020.

Kahneman, Daniel. *Thinking, Fast and Slow*. Farrar, Straus and Giroux. 2011.

Chapter 5

Brueggeman, Walter. *The Prophetic Imagination*. Fortress Press. 2018.

Watts, Alan. *Beyond Theology: The Art of Godmanship*. Vintage Books. 1964.

Chapter 6

Szent-Gyoergyi, Albert. "The Drive in Living Matter to Perfect Itself." *Synthesis 1,* 14-26, 1974.

Haberman, Clyde. "Look Up." *New York Times. December 28, 2019.*

Watson, Galadrial, "Awe Struck." *Discover Magazine.* June, 2020.

Chapter 7

Brooks, David. *The Second Mountain: The Quest for a Moral Life*. Random House. 2019

Chapter 8

Wallace-Wells, David. *The Uninhabitable Earth*. 2019.

Roberts, Jane. "Side-stepping Climate Change, Trump Touts Trillion Trees Plan." *Undark,* February 7, 2020.

Drum, Kevin. "Carbon Emissions Are Up in 2019 Yet Again." *Mother Jones.* December, 4, 2019.

Krugman, Paul. "The Party That Ruined the Planet." *New York Times.* December 12, 2019.

Traub, James. "Our Pursuit of Happiness Is Killing the Planet." *New York Times.* March 6, 2020.

Howell, William Dean. *A Traveler from Altruria.* Harper Publishing House. 1894.

Tabuchi, Hiroko. "A Trump Insider Embeds Climate Denial In Scientific Research. *New York Times. March 2, 2020.*

Mecham, Jon. *The Soul of America: The Battle for Or Better Angels.* Random House. 2018

Chapter 9

Chappell, David. *Will War Ever End? A Soldier's Vision of Peace for the 21st Century. Easton Studio Press. 2009.*

Friedman, Thomas. "With the Coronavirus, It's Again Trump vs. Mother Nature." *New York Times.* March 31, 2020.

Kelly, Scott. "I Spent a Year in Space and I Have Tips on Isolation to Share." *New York Times.* March 21, 2020.